IB PHYSICS REVISION GUIDE

IB PHYSICS REVISION GUIDE

Tim Hoffmann

ZOUEV PUBLISHING

This book is printed on acid-free paper.

Copyright © 2013 Tim Hoffmann. All rights reserved.

No part of this book may be used or reproduced in any manner whatsoever without written permission, except in the case of brief quotations embodied in critical articles or reviews.

Published 2013

Printed by Lightning Source

ISBN 978-0-9560873-9-3, paperback.

TABLE OF CONTENTS

TABLE OF CONTENTS	1
INTRODUCTION	4
HOW TO USE	4
ABOUT THE AUTHOR	4
REFERENCES	5
CONSTANTS	5
SI BASE UNITS	5
SI DERIVED UNITS	5
SI MULTIPLIERS	6
CONVERSIONS	6
THE GREEK APLHABET	6
ELECTROMAGNETIC RADIATION SPECTRUM	7
ELECTRIC CIRCUIT SYMBOLS	7
PERIODIC TABLE OF ELEMENTS	8
LAWS	10
Standard Level Syllabus	10
TOPIC 2: Mechanics	10
TOPIC 4: Oscillations & Waves	10
TOPIC 5: Electric Currents	10
TOPIC 6: Fields & Forces	11
TOPIC 7: Atomic & Nuclear Physics	11
TOPIC 8: Energy, Power & Climate Change	11
Higher Level Syllabus	12
TOPIC 9: Fields & Forces	12
TOPIC 10: Thermal Physics	12
TOPIC 11: Wave Phenomena	12
TOPIC 12: Electromagnetic Induction	12
TOPIC 13: Quantum Physics and Nuclear Physics	12
FORMULAE	14
Standard Level Syllabus	14
TOPIC 2: Mechanics	14
TOPIC 3: Thermal Physics	17
TOPIC 4: Oscillations & Waves	19
TOPIC 5: Electric Currents	20
TOPIC 6: Fields & Forces	21
TOPIC 7: Atomic & Nuclear Physics	23
TOPIC 8: Energy, Power & Climate Change	23
TOPIC 9: Motions in Fields	24
TOPIC 10: Thermal Physics	26

TOPIC 11: Wave Phenomena	27
TOPIC 12: Electromagnetic Induction	28
TOPIC 13: Quantum Physics & Nuclear Physics	29
TOPIC 14: Digital Technology	30

DEFINITIONS 32

Core Syllabus — 32

TOPIC 1: Physics & Physical Measurement 32
- 1.2 Measurements & Uncertainties 32

TOPIC 2: Mechanics 34
- 2.1 Kinematics 34
- 2.2 Forces & Dynamics 34
- 2.3: Work, Energy & Power 35
- 2.4: Uniform Circular Motion 35

TOPIC 3: Thermal Physics 36
- 3.2 Thermal Properties of Matter 36

TOPIC 4. Oscillations & Waves 38
- 4.4 Wave Characteristics 39
- 4.5 Wave Properties 40

TOPIC 5. Electric Currents 41

TOPIC 6: Fields & Forces 43

TOPIC 7: Atomic & Nuclear Physics 44
- 7.2 Radioactive Decay 44
- 7.3 Nuclear Reactions, Fission & Fusion 45

TOPIC 8: Energy, Power & Climate Change 46
- 8.5 Greenhouse Effect 47

Higher Level Syllabus — 49

TOPIC 9: Motion in Fields 49
- 9.2 Gravitation Field, Potential & Energy 49
- 9.3: Electric Field, Potential & Energy 49

TOPIC 10: Thermal Physics 51
- 10.2 Processes 51

TOPIC 11: Wave Phenomena 52
- 11.2 Doppler Effect 52
- 11.3 Diffraction 53

TOPIC 12: Electromagnetic Induction 54

TOPIC 13: Quantum Physics and Nuclear Physics 55

TOPIC 14: Digital Technology 56

DERIVATIONS 58

Standard Level Syllabus — 58

TOPIC 3: Thermal Physics 58

TOPIC 4: Waves 59

TOPIC 5: Electricity & Magnetism 59

- Higher Level Syllabus 61
  - TOPIC 10: Thermal Physics 61
  - TOPIC 11: Wave Phenomena 62
  - TOPIC 12: Electromagnetism 64
  - TOPIC 13: Quantum & Nuclear Physics 65
- METHOD & REVISION TECHNIQUES 67
  - Plotting/Graphing Techniques 67
  - Trigonometry 67
  - Mathematical Techniques 68
  - Method & Simplifications 69
  - Exam Technique 70

INTRODUCTION

This book is intended to act as a reference guide rather than a textbook. It is structured such that the syllabus is broken down into the relevant sections so that if you are searching for a particular word or formula, you know exactly where to look. Whilst taking the IB myself, and tutoring the subject to students, I realised there wasn't a guide like this to make referencing easier and I hope you find it as useful as I think I may have several years ago.

Please note that this guide only acts as a reference for the core Standard & Higher level syllabi. Options are not included.

I hope you find this book useful, even if you only use it as a data booklet or dictionary. Hopefully, you will appreciate what each formula is trying to help you find and the variables that are required. Ensure that as you learn a topic, you understand the definitions and formulae, rather than simply memorising them. The beauty of the physics course is that it's set out in a strategic manner, enabling you to link the various topics in a logical manner.

In a way, by using other guides whilst studying and teaching, I feel I've assembled all the best bits in an accessible 'go-to' guide to pick up nuggets of useful information, rather than wordy explanations like a textbook.

I recommend that you read through this book thoroughly once so that you know where everything is located, and from then on you can use it whilst doing homework, past papers or revising. The book has been structured around the most recent IB syllabus, but from my experiences of A-level and related courses, it may act as a useful reference for various fields of pre-university study.

Good luck!

HOW TO USE

This guide is broken down into sections, and this is done in part so as not to repeat itself. Therefore, when you look up a definition in the DICTIONARY section, there may be some text informing you to refer to the FORMULAE section. This may be because a certain word is inherently linked to a specific formula, or vice versa. Read through the guide at ease in order to see how it has all been pieced together.

Hopefully you will realise the structure is easiest not as a skim-through syllabus notes, but as a reference guide when you're doing tests or exams. It may be useful to read through the laws and definitions several times, as these need memorizing.

Additionally, there is lots of blank space left for you to make your own markings. The book has been done in black & white because it is often easier to annotate and use colour only where you find it most useful, and there is plenty of space to annotate.

ABOUT THE AUTHOR

Tim Hoffmann completed the IB in 2005, obtaining 44 points, with a 7 in HL Physics. Tim furthered his studies at St Edmund Hall, Oxford University, receiving a Masters of Engineering Science in 2009. After providing several years of tutoring experience for IBDP and MYP students, he set up a tutoring agency catering for IB students in the UK. Elite IB has subsequently gone from strength to strength and now caters for students around the world, with very close ties to several UK IB schools. Elite IB offer individual tuition as well as online tuition, summer revision courses, and university admission mentoring.

See www.eliteib.co.uk for details.

For feedback or suggestions, please contact book@eliteib.co.uk

REFERENCES

A large amount of this section is available in your data booklet, but in order to help while studying, it's repeated here. Later sections of this guide will refer back to these values and symbols. You do NOT need to memorise them, but should understand their units and relevance.

CONSTANTS

Quantity	Symbol	Value (approximate)
Gravitational acceleration (near earth)	g	$9.81 ms^{-2}$
Gravitational constant	G	$6.67 \times 10^{-11} Nm^2kg^{-1}$
Avagadro's constant	N_A	$6.02 \times 10^{23} mol^{-1}$
Gas constant	R	$8.31 JK^{-1}mol^{-1}$
Boltzmann's constant	k	$1.38 \times 10^{-23} JK^{-1}$
Stefan-Boltzmann's constant	σ	$5.67 \times 10^{-8} Wm^{-2}K^{-1}$
Coulomb constant	k	$8.99 \times 10^9 Nm^2C^{-2}$
Permittivity of free space	ε_0	$8.85 \times 10^{-12} C^2N^{-1}m^{-2}$
Permeability of free space	μ_0	$4\pi 10^{-7} TmA^{-1}$
Speed of light in a vacuum	c	$3.00 \times 10^8 ms^{-1}$
Planck's constant	h	$6.63 \times 10^{-34} Js$
Elementary charge (charge on an electron or proton)	e	$1.60 \times 10^{-19} C$
Electron rest mass	m_e	$9.11 * 10^{-31} kg = 0.00059u = 0.511 MeVe^{-2}$
Proton rest mass	m_p	$1.673 * 10^{-27} kg = 1.00726u = 938 MeVe^{-2}$
Neutron rest mass	m_n	$1.675 * 10^{-27} kg = 1.008665u = 940 MeVe^{-2}$
Unified atomic mass unit	u	$1.661 * 10^{-27} kg = 931.5 MeVe^{-2}$

SI BASE UNITS

Quantity	Name	Symbol/SI Unit
Length	Metre	m
Mass	Kilogram	kg
Time	Second	s
Current	Ampere	A
Temperature	Kelvin	K
Amount of substance	Mole	mol
Luminous Intensity	Candela	Cd

SI DERIVED UNITS

Derived Unit	Symbol	SI Base Unit
Newton	N	$kgms^{-2}$
Pascal	Pa	$kgm^{-1}s^{-2}$
Hertz	Hz	s^{-1}
Joule	J	kgm^2s^{-2}
Watt	W	kgm^2s^{-3}
Coulomb	C	As
Volt	V	$kgm^2s^{-3}A^{-1}$
Ohm	Ω	$kgm^2s^{-3}A^{-2}$
Weber	Wb	$kgm^2s^{-2}A^{-1}$
Tesla	T	$kgs^{-2}A^{-1}$
Becquerel	Bq	s^{-1}

SI MULTIPLIERS

PREFIX (symbol)	VALUE
Tera (T)	10^{12}
Giga (G)	10^{9}
Mega (M)	10^{6}
Kilo (k)	10^{3}
Hecto (h)	10^{2}
Deca (da)	10^{1}
Deci (d)	10^{-1}
Centi (c)	10^{-2}
Milli (m)	10^{-3}
Micro (μ)	10^{-6}
Nano (n)	10^{-9}
Pico (p)	10^{-12}
Femto (f)	10^{-15}

CONVERSIONS

There are several conversions which are included in the *FORMULAE* section rather than here, where these are the most common. Unlike the formula, it IS useful to memorise these as they come up frequently, and will seed up how quickly you can respond to an exam question.

$1\ radian = \dfrac{180°}{\pi}$

1 kilowatthour $kwH^{-1} = 3.60 * 10^{6} J$

$1 atm = 760 mmHg = 33.9 ftwater^{-1}$ ($at\ 0°C$)

1 calorie = 4.18 Joules

1 Joule = 1 Newton metre = 1 Wattsecond

1 electronvolt = $1.602 * 10^{-19}$ Joules

1 bar = 10^5 pascals per square metre = 0.986923 atmospheres

THE GREEK APLHABET

Letter			Letter		
Capital	Lower-case	Name	Capital	Lower-case	Name
A	α	Alpha	N	ν	Nu
B	β	Beta	Ξ	ξ	Xi
Γ	γ	Gamma	O	o	Omicron
Δ	δ	Delta	Π	π	Pi
E	ε	Epsilon	P	ρ	Rho
Z	ζ	Zeta	Σ	σ	Sigma
H	η	Eta	T	τ	Tau
Θ	θ	Theta	Y	υ	Upsilon
I	ι	Iota	Φ	φ	Phi
K	κ	Kappa	X	χ	Chi
Λ	λ	Lambda	Ψ	ψ	Psi
M	μ	Mu	Ω	ω	Omega

ELECTROMAGNETIC RADIATION SPECTRUM

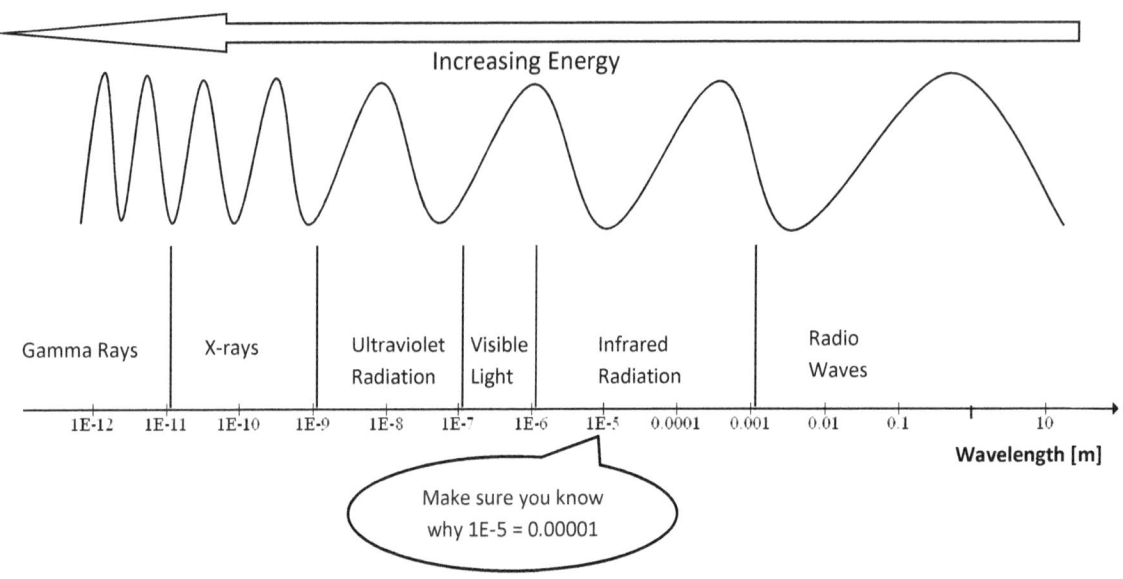

ELECTRIC CIRCUIT SYMBOLS

The symbols below are the only ones required for IB Physics exams, and it is useful to memorise these. One useful hint is that for an AC motor, the 'ac supply' symbol will be replaced without the *swish* inside, to show that it will **generate** an AC signal between the anode and cathode rather than supply one.

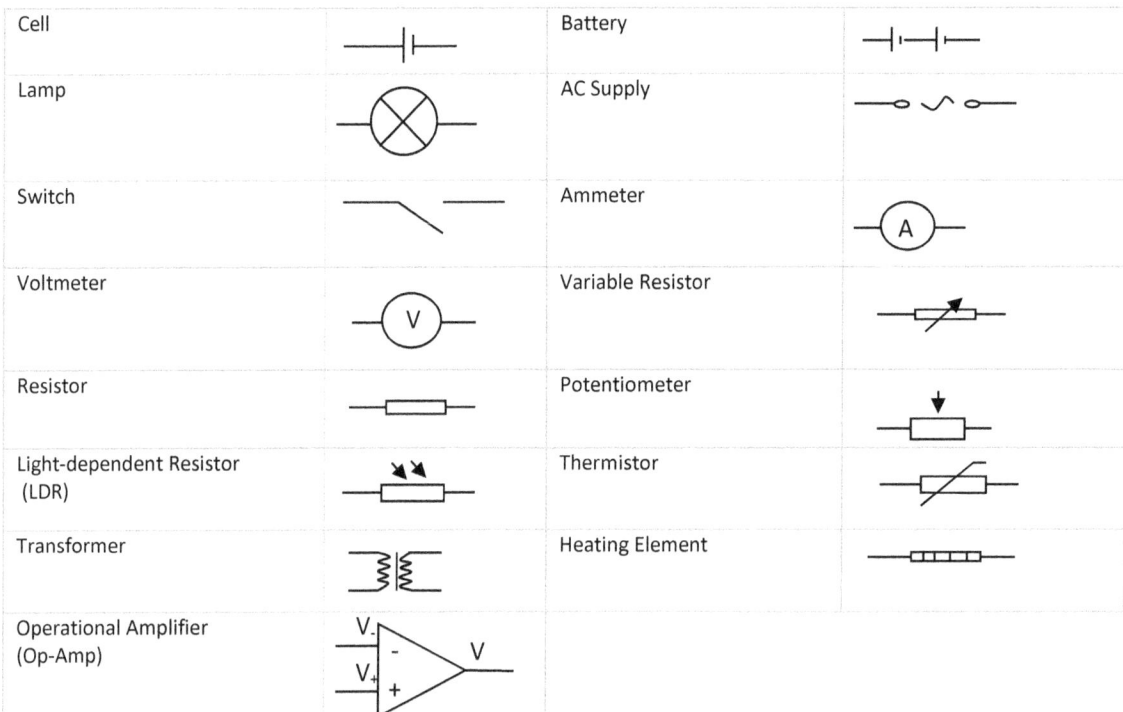

PERIODIC TABLE OF ELEMENTS

1 H Hydrogen 1.00794																	2 He Helium 4.003
3 Li Lithium 6.941	4 Be Beryllium 9.012182											5 B Boron 10.811	6 C Carbon 12.0107	7 N Nitrogen 14.00674	8 O Oxygen 15.9994	9 F Fluorine 18.9984032	10 Ne Neon 20.1797
11 Na Sodium 22.989770	12 Mg Magnesium 24.3050											13 Al Aluminum 26.981538	14 Si Silicon 28.0855	15 P Phosphorus 30.973761	16 S Sulfur 32.066	17 Cl Chlorine 35.4527	18 Ar Argon 39.948
19 K Potassium 39.0983	20 Ca Calcium 40.078	21 Sc Scandium 44.955910	22 Ti Titanium 47.867	23 V Vanadium 50.9415	24 Cr Chromium 51.9961	25 Mn Manganese 54.938049	26 Fe Iron 55.845	27 Co Cobalt 58.933200	28 Ni Nickel 58.6934	29 Cu Copper 63.546	30 Zn Zinc 65.39	31 Ga Gallium 69.723	32 Ge Germanium 72.61	33 As Arsenic 74.92160	34 Se Selenium 78.96	35 Br Bromine 79.904	36 Kr Krypton 83.80
37 Rb Rubidium 85.4678	38 Sr Strontium 87.62	39 Y Yttrium 88.90585	40 Zr Zirconium 91.224	41 Nb Niobium 92.90638	42 Mo Molybdenum 95.94	43 Tc Technetium (98)	44 Ru Ruthenium 101.07	45 Rh Rhodium 102.90550	46 Pd Palladium 106.42	47 Ag Silver 107.8682	48 Cd Cadmium 112.411	49 In Indium 114.818	50 Sn Tin 118.710	51 Sb Antimony 121.760	52 Te Tellurium 127.60	53 I Iodine 126.90447	54 Xe Xenon 131.29
55 Cs Cesium 132.90545	56 Ba Barium 137.327	57 La Lanthanum 138.9055	72 Hf Hafnium 178.49	73 Ta Tantalum 180.9479	74 W Tungsten 183.84	75 Re Rhenium 186.207	76 Os Osmium 190.23	77 Ir Iridium 192.217	78 Pt Platinum 195.078	79 Au Gold 196.96655	80 Hg Mercury 200.59	81 Tl Thallium 204.3833	82 Pb Lead 207.2	83 Bi Bismuth 208.98038	84 Po Polonium (209)	85 At Astatine (210)	86 Rn Radon (222)
87 Fr Francium (223)	88 Ra Radium (226)	89 Ac Actinium (227)	104 Rf Rutherfordium (261)	105 Db Dubnium (262)	106 Sg Seaborgium (263)	107 Bh Bohrium (262)	108 Hs Hassium (265)	109 Mt Meitnerium (266)	110 (269)	111 (272)	112 (277)	113	114				

58 Ce Cerium 140.116	59 Pr Praseodymium 140.90765	60 Nd Neodymium 144.24	61 Pm Promethium (145)	62 Sm Samarium 150.36	63 Eu Europium 151.964	64 Gd Gadolinium 157.25	65 Tb Terbium 158.92534	66 Dy Dysprosium 162.50	67 Ho Holmium 164.93032	68 Er Erbium 167.26	69 Tm Thulium 168.93421	70 Yb Ytterbium 173.04	71 Lu Lutetium 174.967
90 Th Thorium 232.0381	91 Pa Protactinium 231.03588	92 U Uranium 238.0289	93 Np Neptunium (237)	94 Pu Plutonium (244)	95 Am Americium (243)	96 Cm Curium (247)	97 Bk Berkelium (247)	98 Cf Californium (251)	99 Es Einsteinium (252)	100 Fm Fermium (257)	101 Md Mendelevium (258)	102 No Nobelium (259)	103 Lr Lawrencium (262)

NOTES

LAWS

The IB syllabus outlines the importance of several laws upon which many of the topics are based. If you really understand the underlying principles of the law, you will spot examples of it in P1 questions immediately, removing the need for any (serious) thought! It's also helpful to memorise these definitions for use in P2 questions, where the first part often asks you to simply state the law. Try and associate a law with a picture in your head – most of these laws can be easily visualised as real-life examples.

Standard Level Syllabus

TOPIC 2: Mechanics

× Newton's First Law of Motion: a.k.a 'The definition of force' – Every object will continue in a state of rest or uniform motion in a straight line unless an external force acts on it.

× Newton's Second Law of Motion:. $F = m * a = \frac{\Delta p}{\Delta t}$. The rate of change of momentum of an object is directly proportional to the force acting on it, and will take place in a straight line in which the force acts.

× Law of Conservation of Momentum: In any system where two or more objects act on each other, their total linear momentum in any given direction remains constant, unless an external force acts in that direction.

× Newton's 3rd Law of Motion: $\sum F_{net} = 0$. If two bodies interact, the force exerted by body A upon body B will be equal and opposite to the force exerted by B upon A. They are simultaneous, equal in magnitude and opposite in direction. "Every action has an equal and opposite reaction."

× Principle of Conservation of Energy: Energy cannot be created nor destroyed, but it can be changed from one form into another.

TOPIC 4: Oscillations & Waves

× Snell's Law: a.k.a 'Law of refraction', $\frac{n_r}{n_i} = \frac{\sin(\theta_i)}{\sin(\theta_r)} = \frac{v_i}{v_r}$, where subscript *i* refers to incident, *r* refers to refracted, *n* are the indices and *v* are the velocities in the respective media. This applies for a constant frequency, and thus the law is a consequence of a change in speed of a wave. See DERIVATIONS for a graphical derivation.

TOPIC 5: Electric Currents

× Ohm's Law: The current flowing through a conductor is directly proportional to the potential difference across its ends, provided the temperature and other physical conditions remain constant. This law is only truly accurate for metals.

× Fleming's Left-Hand Rule: a.k.a 'Motor Principle'. Use three fingers for field (Index finger), current (Middle finger) and force (Thumb) in reference to a conductor moving through a magnetic field, in order to find the direction of the three variables. See page 12.

× Fleming's Right-Hand Rule: a.k.a 'Dynamo Principle'. To find the direction of the induced current, thumb points in direction of flow of current, fingers point in direction of induced current/field.

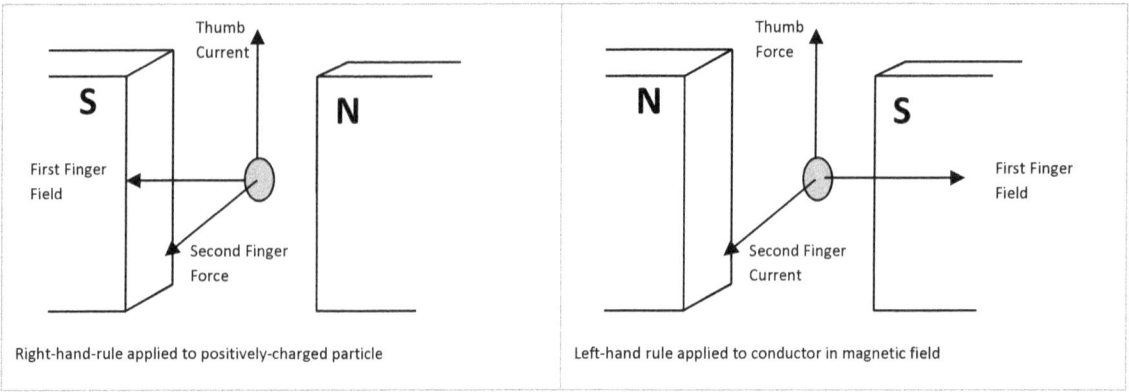

Right-hand-rule applied to positively-charged particle | Left-hand rule applied to conductor in magnetic field

TOPIC 6: Fields & Forces

× Newton's Law of Gravitation: $F = \frac{k(m_1 M_2)}{r^2}$. Best seen in *FORMULAE*. Law applying to two point masses stating a version of Newton's 3rd law that they exert equal and opposite forces on each other, always in an attractive manner.

× Law of Conservation of Charge: The total net charge on a system is constant.

× Coulomb's Law: $F = \frac{q_1 * q_2}{4 * \pi * \varepsilon * r^2}$. Best seen in *FORMULAE*. Applies for two point charges, stating a version of Newton's 3rd law that they exert equal and opposite charges on each other, which may be either attractive or repulsive.

TOPIC 7: Atomic & Nuclear Physics

× Einstein Mass-Energy Equivalence Relationship: $E = mc^2$, see *FORMULAE*. A body of mass 'm' has associated energy 'E', and a quantity of energy 'E' has associated mass 'm'. A law stating that the two quantities are interchangeable.

TOPIC 8: Energy, Power & Climate Change

× Stefan-Boltzman Law: $M_e = \sigma T^4$, Links the radiant flux per unit area emitted by a black body to its absolute (Kelvin) temperature.

Higher Level Syllabus

TOPIC 9: Fields & Forces

- Equipotential Law: Equipotentials (for both gravitational and electrical cases) always intersect field lines at 90°, and are spaced so as to reveal the rate at which the field is changing. I.e. equipotentials will get further apart as you move further from the field-creating body.

- Kepler's 3rd Law: "The squares of the times taken to describe their orbits by two planets are proportional to the cubes of the major semi-axes of the orbits." This essentially means that for any mass orbiting a larger mass, the ratio, $\frac{T^2}{R^3}$ will be constant, where T is period of orbit and R is radius of orbit. SEE DERIVATIONS.

TOPIC 10: Thermal Physics

- First Law of Thermodynamics: Essentially a means of stating the principle of conservation of energy. "Heat is a process of energy transfer, and in a closed system the total amount of energy is constant."

- Second Law of Thermodynamics: This law suggests that energy cannot randomly transfer from a region of low temperature to high temperature, and means that the entropy in the universe must always be increasing.

TOPIC 11: Wave Phenomena

- Rayleigh Criterion: For the resolution of two point sources, the inner dark ring of one diffraction pattern should coincide with the centre of the second diffraction pattern.

- Brewster's Law: When light strikes a glass surface at an angle of incidence given by $\tan^{-1}(n)$, (where n is the refractive index) the reflected light is *plane-polarized*. At this angle of incidence, the refracted ray makes an angle 90° with the reflected ray.

- Malus' Law: The transmission of plane-polarized light from a polarizer through a second identical polarizer varies as $\cos^2(\theta)$, where θ is the angle between the transmission axes and the two polarizers. SEE FORMULAE.

TOPIC 12: Electromagnetic Induction

- Faraday's Law: For a current-carrying wire in a magnetic field, the time-varying *flux linkage* in the wire-coil induces an electromotive force (e.m.f) in the wire.

- Lenz' Law: The direction of the induced e.m.f in a conductor is such that if an induced current were able to flow, it would oppose the change which caused it.

TOPIC 13: Quantum Physics and Nuclear Physics

- Radioactive Decay Law: A law stating that a radioactive substance with unstable nuclei can only be predicted as to how probable the decay will be, rather than an exact relationship. This is based on an exponential equation linking the number of unstable nuclei with the predicted half-life. SEE FORMULAE.

- Heisenberg Uncertainty Principle: A principle which states that pairs of physical properties (e.g. position and momentum), cannot both be known to a particular precision. For example, if we know a body's position to a more precise value, we will know its momentum to a less precise value. It is closely linked in with wave-particle duality and the inherent lack of clarity.

NOTES

FORMULAE

The data booklet is your best friend in the exam. Read it over and over to establish a.) what each equation finds, b.) what each variable in the equation stands for and c.) the units of each variable. If you know these, derivations become easier and you'll be able to know straight away which equations you will need. Here is a list of almost ALL the equations you may need, many of which **aren't** offered in the data booklet. Use this guide when doing examples, but try and establish how you can derive these from the equations available to you in the exams!

The variables in [] represent the SI units, which are found in the introduction section. The (v) or (s) after each variable represents *vector* and *scalar* respectively, included where relevant. Note that the data booklet you receive has vectors quantities in **bold**.

Standard Level Syllabus
TOPIC 1: Physics & Physical Measurement

Addition/Subtraction	
$E_{tot} = Error_1 + Error_2$	*When two values are either added or subtracted, their errors are always added.*
Multiplication/Division	
$E_{tot} = \dfrac{Error_1}{Value_1} + \dfrac{Error_2}{Value_2}$	*When two values are either multiplied or divided, their fractional errors are added. Also, if something is squared, this means it is (itself)x(itself), so its fractional error must be doubled.*

TOPIC 2: Mechanics
These first three equations are broken up more simply to see the general style of presentation.

Variables	Formula	Units	Result
$speed = \dfrac{distance}{time}$	$v = \dfrac{d}{t}$	ms^{-1}	Scalar
$velocity = \dfrac{displacement}{time}$	$v = \dfrac{s}{t}$	ms^{-1}	Vector
$acceleration = \dfrac{change\ in\ velocity}{time}$	$v = \dfrac{\Delta v}{\Delta t}$	ms^{-2}	Vector

KINEMATICS	
$v = u + at$ $s = \dfrac{u+v}{2} * t$ $v^2 = u^2 + 2as$ $s = ut + \dfrac{1}{2}(at^2)$ $s = vt - \dfrac{1}{2}(at^2)$	u = Initial velocity [ms^{-1}] (v) v = Final velocity [ms^{-1}] (v) a = Acceleration [ms^{-2}] (v) t = Time taken [s] (s) s = Distance travelled [m] (s) For all of these equations, consider balancing the units on either side. Note: These formulae assume constant acceleration

$a = \dfrac{v-u}{t}$ $v_{average} = \dfrac{u+v}{2}$	u = Initial velocity [ms^{-1}] (v) v = Final velocity [ms^{-1}] (v) a = Acceleration [ms^{-2}] (v) t = Time taken [s] (s) s = Distance travelled [m] (s)
SPRING/DAMPER	
$F = -kx$	k = Spring constant [Nm^{-1}] (s) x = Displacement [m] (v) F = Force [N] (v) *Note: Right hand side (RHS) is negative because the force always acts to **restore** the spring to its equilibrium position*
FORCE/MOMENTUM	
$F = \dfrac{\Delta p}{\Delta t}$	F = Force [N] (v) Δp = Change in momentum [$kgms^{-1}$] (v) Δt = Change in time [s] (s)
$F = ma$	F = Force [N] (v) m = Mass [kg] (s) a = Acceleration [ms^{-2}] (v)
$p = mv$	p = Momentum [$kgms^{-1}$] (v) m = Mass [kg] (s) v = Velocity [ms^{-1}] (v)
WORK/POWER/ENERGY	
$W = Fd$ $W = Fd\cos(\theta)$ **When lifting something:** $W = mgh$	W = Work done [J] or [Nm^2] (s) F = Force [N] (v) d = Displacement [m] (v) θ = Angle between force and direction body moves in m = Mass [kg] (s) g = Gravitational acceleration [ms^{-2}] (v) h = Height, equivalent to displacement [m] (s) *This an important equation which will often be the first step in derivations, and you must remember that while Work and Energy are scalars, Force and Displacement are vectors, so **no work is done if force and displacement are at 90°**.*
$P = \dfrac{W}{t}$	P = Power [$Nm^2 s^{-1}$] or [W] (s) W = Work done [J] or [Nm^2] (s) t = Time [s] (s) Power = Work Done/ Time Taken

$PE_k = KE = \frac{1}{2}mv^2 = \frac{p^2}{2v}$	E_k = Kinetic Energy [Nm^2] (s) m = Mass [kg] (s) v = Velocity [ms^{-1}] (v) ρ = Momentum [$kgms^{-1}$] (v)
$Power = Fv$	F = Force [N] (v) v = Velocity [ms^{-1}] (v)
$Impulse = F\Delta t = m\Delta v$	F = Force [N] (v) t = Time [s] (s) m = Mass [kg] (s) v = Velocity [ms^{-1}] (v)
WORK DONE ON/BY A SPRING	
$W = \frac{1}{2}kx^2$	W = Work done on/by spring [J] (s) k = Spring constant [Nm^{-1}] (s) x = Extension of spring from equilibrium position [m] (v) *Force [N] vs Displacement [m] graph: W = Area*
CIRCULAR MOTION	
$a_{centr} = \frac{v^2}{r}$ $F_{centr} = \frac{mv^2}{r}$	a_{centr} = Centripetal acceleration [ms^{-2}] (v) F_{centr} = Centripetal force [N] (v) v = Velocity [ms^{-1}] (v) m = Mass [kg] (s) r = Radius [m] (s) *Remember, **no work is done** in **uniform** circular motion.*

TOPIC 3: Thermal Physics

TEMPERATURE & INTERNAL ENERGY	
$T(K) = t(C) + 273$ 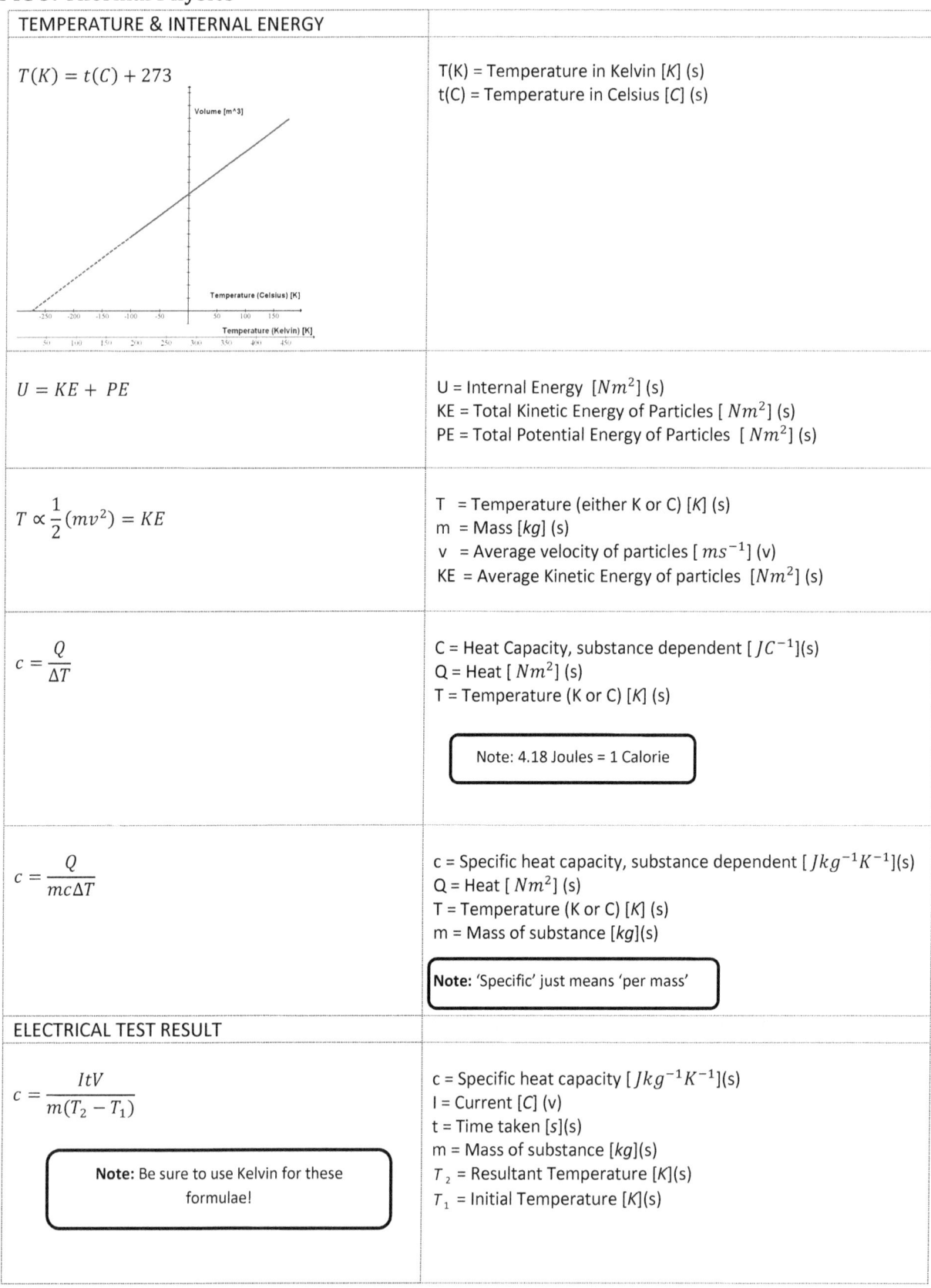	T(K) = Temperature in Kelvin [K] (s) t(C) = Temperature in Celsius [C] (s)
$U = KE + PE$	U = Internal Energy $[Nm^2]$ (s) KE = Total Kinetic Energy of Particles $[Nm^2]$ (s) PE = Total Potential Energy of Particles $[Nm^2]$ (s)
$T \propto \frac{1}{2}(mv^2) = KE$	T = Temperature (either K or C) [K] (s) m = Mass [kg] (s) v = Average velocity of particles $[ms^{-1}]$ (v) KE = Average Kinetic Energy of particles $[Nm^2]$ (s)
$c = \frac{Q}{\Delta T}$	C = Heat Capacity, substance dependent $[JC^{-1}]$(s) Q = Heat $[Nm^2]$ (s) T = Temperature (K or C) [K] (s) Note: 4.18 Joules = 1 Calorie
$c = \frac{Q}{mc\Delta T}$	c = Specific heat capacity, substance dependent $[Jkg^{-1}K^{-1}]$(s) Q = Heat $[Nm^2]$ (s) T = Temperature (K or C) [K] (s) m = Mass of substance [kg](s) Note: 'Specific' just means 'per mass'
ELECTRICAL TEST RESULT	
$c = \frac{ItV}{m(T_2 - T_1)}$ Note: Be sure to use Kelvin for these formulae!	c = Specific heat capacity $[Jkg^{-1}K^{-1}]$(s) I = Current [C] (v) t = Time taken [s](s) m = Mass of substance [kg](s) T_2 = Resultant Temperature [K](s) T_1 = Initial Temperature [K](s)

MIXTURE TEST RESULT	
$m_a c_a (T_a - T_{\max}) = m_b c_b (T_{\max} - T_b)$	a = Substance A b = Substance B Subscript max = Resultant Substance M = Mass [kg](s) c = Specific heat capacity [$Jkg^{-1}K^{-1}$](s) T = Temperature [K](s)
$l = \dfrac{Q}{m}$	l = Latent heat capacity [Jkg^{-1}](s) Q = Heat [Nm^2] (s) m = Mass of substance [kg](s)
$l = \dfrac{ItV}{m_1 - m_2}$	c = Specific heat capacity [$Jkg^{-1}K^{-1}$](s) I = Current [C](v) t = Time taken [s](s)
$m_{wat} c_{wat}(T_{wat} - T_{mix}) = m_{ice} l_{ice} + m_{ice} c_{wat} T_{mix}$	wat = Substance Water ice = Substance Ice mix = Resultant Mixture m = Mass of substance [kg](s) l = Latent heat capacity [Jkg^{-1}](s) c = Specific heat capacity [$Jkg^{-1}K^{-1}$](s)
GAS LAWS	
$PV = nRT$	P = Pressure [Pa](s) V = Volume [m^3](s) n = Number of moles (s) R = Gas Constant [$K^{-1}mol^{-1}$](s) T = Temperature [K](s)
$P = \dfrac{F}{A}$	P = Pressure [Pa](v) F = Force [N] (v) A = Area [m^2](s)
$n = \dfrac{N}{N_A}$	n = Number of moles N = Atomic number (i.e. number of nucleons) N_A = Avagadro's Constant [mol^{-1}](s)
HEAT TRANSFER	
$\dfrac{Q}{\Delta t} = kA \dfrac{T}{\Delta x}$	Q = Heat [Nm^2] (s) t = Time taken [s](s) A = Area of contact [m^2](s) T = Temperature [K](s) x = Distance in contact [m](s)

TOPIC 4: Oscillations & Waves

SIMPLE HARMONIC MOTION	
$\omega = \dfrac{2\pi}{T}$	ω = Angular velocity [s^{-1}](v) T = Period of rotation [s](s)
$x = x_o \sin(\omega t), x = x_o \cos(\omega t)$ $v = v_o \sin(\omega t), v = v_o \cos(\omega t)$ $v = \pm \omega \sqrt{x_o^2 - x^2}$	x = Displacement of particle [m](v) v = Velocity of particle [ms^{-1}](v) x_o = Maximum amplitude of displacement [m](s) v_o = Maximum amplitude of velocity [m](s)
$E_k = \dfrac{1}{2}(m\omega^2 x_o^2)$ $E_T = \dfrac{1}{2}(m\omega^2 x_o^2)$ $T = \dfrac{1}{f}$	E_k = Kinetic Energy [Nm^2] (s) T = Period of rotation [s](s) f = Frequency [Hz](s) m = Mass of particle [kg](s) ω = Angular velocity [s^{-1}](v)

[Graph: Displacement [m] vs Time [s], showing sinusoidal wave with period T labeled]

[Graph: Displacement [m] vs Distance [m], showing sinusoidal wave with wavelength λ labeled]

$v = \dfrac{distance}{time} = \dfrac{\lambda}{T} = f\lambda$	v = Wave velocity [ms^{-1}] (v) f = Wave Frequency [Hz](s) λ = Wavelength [m](s) T = Period of oscillation [s](s)
REFRACTION/DIFFRACTION	
$\dfrac{n_1}{n_2} = \dfrac{\sin(\theta_1)}{\sin(\theta_2)} = \dfrac{v_1}{v_2}$	1 = Subscript of medium 1 2 = Subscript of medium 2 n = Refractive index (ratio) θ = Angle of refraction relative to normal (angle) v = Wave velocity [ms^{-1}] (v)
$n\lambda = path\ difference$ OR $n + \left(\dfrac{1}{2}\right)\lambda = path\ difference$	n = Number of wavelengths λ = Wavelength [m](s)

SPRING/DAMPER		
$F = -kx$	k = Spring constant $[Nm^{-1}]$ (s) x = Displacement $[m]$ (v) F = Force $[N]$ (v)	**Note:** Right hand side is negative because the force always acts to restore the spring to its equilibrium position.

TOPIC 5: Electric Currents

CHARGE	
$p.d = path\ difference\ between\ two\ points.$ $= \dfrac{energy\ difference}{charge\ moved}$	p.d = Potential difference $[JC^{-1}]$ (s) $1\ Volt = 1\ JC^{-1}$
$Energy\ gained = p.d * charge$	
$\dfrac{1}{2}mv^2 = Ve$	m = Mass $[kg]$ (s) v = Velocity $[ms^{-1}]$ (v) V = Voltage $[JC^{-1}]$ (s) e = Charge on an electron $[C]$ (v)
CHARGE MOVING IN A FIELD	
$v = \sqrt{\dfrac{2Eqd}{m}}$	v = Velocity $[ms^{-1}]$ (v) E = Electric Field Strength $[NC^{-1}]$ (v) m = Mass of charge $[kg]$ (s) d = Displacement in direction of field by charge $[m]$ (v)
CORE EQUATIONS	
$I = \dfrac{Charge\ Flowed}{Time\ Taken} = \dfrac{Q}{t}$	I = Current $[A]$ (v) Q = Charge $[C]$ (v) t = time $[s]$ (s)
$R = \dfrac{V}{I}$	R = Resistance $[\Omega]$ (s) V = Voltage $[V]$ (s) I = Current $[C]$ (s)
$R = \dfrac{\rho L}{A}$	R = Resistance $[\Omega]$ (s) ρ = Resistivity $[\Omega m]$ (s) L = Length $[m]$ (s) A = Cross-sectional area $[m^2]$ (s)
POWER	
$P = VI = \dfrac{V^2}{R} = I^2 R$	P = Power $[Nm^2 s^{-1}]$ or $[W]$ (s) V = Voltage $[V]$ (v) I = Current $[A]$ (v)

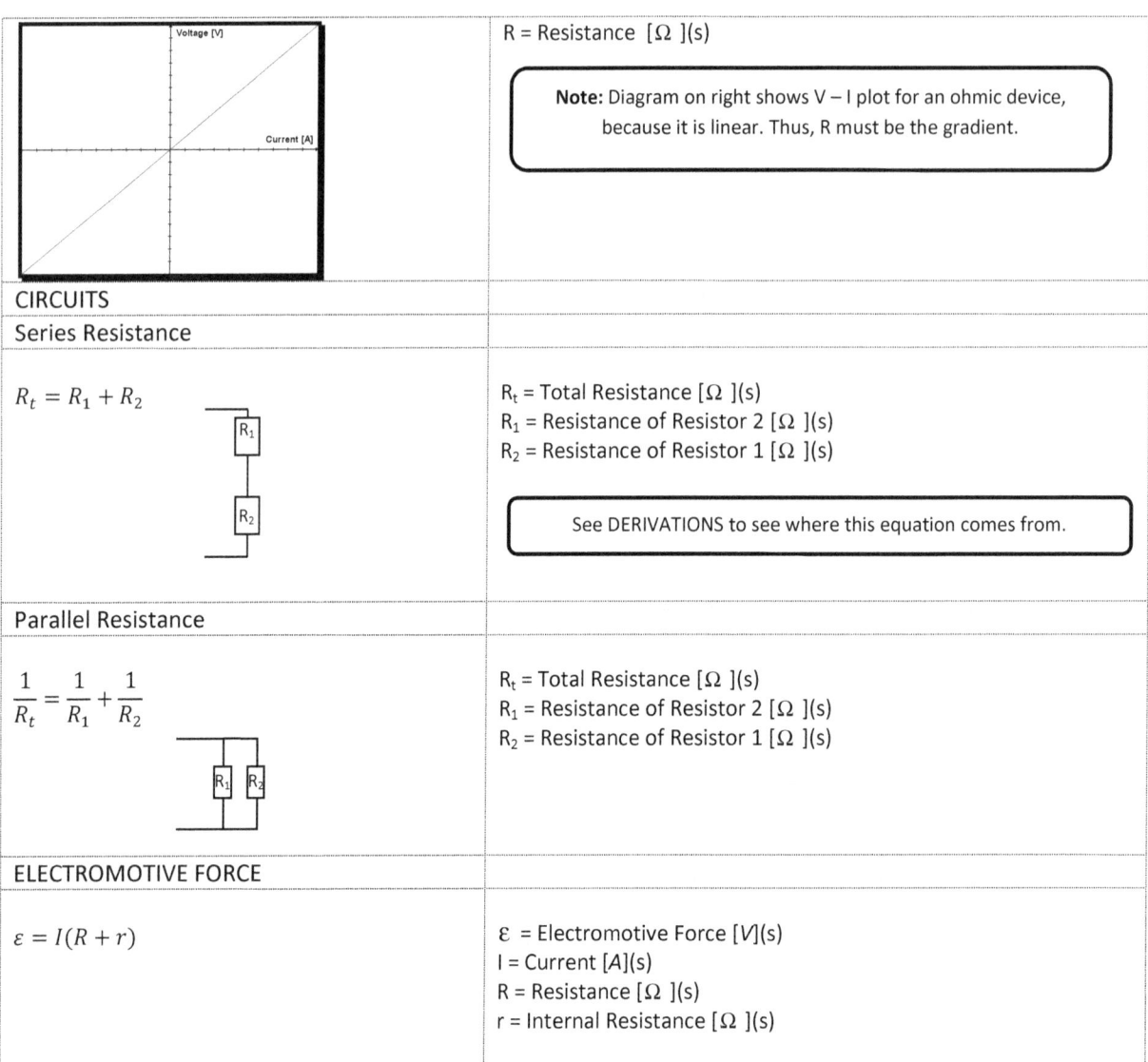	R = Resistance [Ω](s) **Note:** Diagram on right shows V – I plot for an ohmic device, because it is linear. Thus, R must be the gradient.
CIRCUITS	
Series Resistance	
$R_t = R_1 + R_2$	R_t = Total Resistance [Ω](s) R_1 = Resistance of Resistor 2 [Ω](s) R_2 = Resistance of Resistor 1 [Ω](s) See DERIVATIONS to see where this equation comes from.
Parallel Resistance	
$\dfrac{1}{R_t} = \dfrac{1}{R_1} + \dfrac{1}{R_2}$	R_t = Total Resistance [Ω](s) R_1 = Resistance of Resistor 2 [Ω](s) R_2 = Resistance of Resistor 1 [Ω](s)
ELECTROMOTIVE FORCE	
$\varepsilon = I(R + r)$	ε = Electromotive Force [V](s) I = Current [A](s) R = Resistance [Ω](s) r = Internal Resistance [Ω](s)

TOPIC 6: Fields & Forces

GRAVITATIONAL FIELDS	*For fields and forces, it is useful to see the similarities between the equations in gravitational and electrical fields, noting that whilst gravitational fields are always positive, electric fields can also be repulsive.*
$F = \dfrac{Gm_1 m_2}{r}$	F = Equal & opposite forces acted by each mass [N](v) G = Gravitational constant [$Nm^2 kg^{-1}$] m_1 = Mass 1 [kg](s) m_2 = Mass 2 [kg](s) r = Distance separating masses [m](s)
$g = \dfrac{F}{m}$	g = Gravitational acceleration or field strength [ms^{-2}] (v) F = Force exerted on mass at given point in field [N](v) m = Test mass [kg](s)

KEPLER'S 3RD LAW	
$\dfrac{r^3}{T^2} = k$	r = Radius of orbit of satellite [m](s) T = Period of rotation of satellite[s](s) k = Constant which holds for all orbiting bodies |■

ENERGY	
$U = KE + PE = -\dfrac{\frac{1}{2}(GMm)}{r}$	U = Total Energy of orbiting Satellite [J](s) KE = Kinetic Energy of Satellite [Nm^2] (s) PE = Potential Energy of Satellite [Nm^2] (s) G = Gravitational Constant [Nm^2kg^{-2}] M = Mass of field-creating body [kg](s) m = Mass of satellite [kg](s) r = Radius of orbit of satellite [m](s)

Energy [J] vs Radius of Orbit [m]: Kinetic Energy, Total Energy, Gravitational Potential Energy

ELECTRIC FIELDS	
$F = \dfrac{kq_1q_2}{r^2}$	F = Equal & opposite forces acted by each charge [N](v) k = Coulomb Constant [Nm^3C^{-2}] q_1 = Charge on body1 [C](v) q_2 = Charge on body2 [C](v) r = Separation between two bodies [m](s)
$E = \dfrac{F}{q}$	E = Electric field strength [ms^{-2}] (v) F = Force felt by test charge in this field [N](v) q = Charge on test charge placed in field [C](v)

CHARGE MOVING THROUGH MAGNETIC FIELD	
$F = Bqv\sin(\theta)$	F = Force felt by charge in magnetic field [N](v) B = Magnetic Field Strength [T](v) q = Charge of body in field [C](v) v = Velocity of moving charge [ms^{-1}] (v) θ = Angle between normal and velocity direction (angle)

CURRENT-CARRYING CONDUCTOR IN MAGNETIC FIELD	
$F = BIL\sin(\theta)$	F = Force felt by conductor in magnetic field [N](v) B = Magnetic Field Strength [T](v) I = Current in conductor [A](v) L = Length of conductor **in** the field [m](s) θ = Angle between normal and conductor (angle)

$B = \dfrac{\mu I}{2\pi r}$	B = Magnetic Field Strength [*T*](v) μ = Material constant [TmA^{-1}] I = Current in wire [*A*](v) r = Separation [*m*](s)
$F_1 = \dfrac{B_2 I_1 I_2}{2\pi r}$	F_1 = Equal & opposite forces acted by each wire [*N*](v) B_2 = Magnetic field created by Wire 2 [*T*](v) I_1 = Current in wire 1 [*A*](v) I_2 = Current in wire 2 [*A*](v) r = Distance separating two wires [*m*](s) Note: This equation leads to the definition of the Ampere

TOPIC 7: Atomic & Nuclear Physics

CHEMICAL SYMBOL	
X_Z^A	X = Chemical symbol A = Mass number = number of nucleons Z = Atomic number = number of protons
MASS-ENERGY RELATION	
$E = mc^2$	E = Energy of photon [*J*](s) m = Mass of photon [*kg*](s) c = Speed of light [ms^{-1}](v)
CONVERSION	
$p_1^1 \longrightarrow n_0^1 + \beta_{+1}^0 + \upsilon$	P_1^1 = positron n_0^1 = neutron β_{+1}^0 = Beta particle υ = Gamma ray. Energy.

TOPIC 8: Energy, Power & Climate Change

$Power = \dfrac{1}{2}(A\rho v^3)$	A = Area of contact [m^2](s) ρ = Density of air [kgm^{-3}](s) v = velocity of incoming air/wind [ms^{-1}](v) g = gravitational acceleration [ms^{-2}] (v)

$Power\ per\ unit\ length = \frac{1}{2}(A^2 \rho g v)$	A = Area of contact [m^2](s) ρ = Density of air [kgm^{-3}](s) v = velocity of incoming air/wind [ms^{-1}](v) g = gravitational acceleration [ms^{-2}] (v)
$I = \frac{Power}{Area}$	I = Incident power [Wm^{-2}] Power = Power emitted by source [W](s) Area = Area of contact [m^2](s)
$Albedo = \frac{Total\ scattered\ power}{Total\ incident\ power}$	
$C_s = \frac{Q}{A\Delta T}$	C_s = Coefficient of heat transfer Q = Heat flux (W)(s) A = Area of contact [m^2](s) T = Temperature [K]
$Power = \sigma A T^4$ $Power = e\sigma A T^4$	P = Emissive power of a black body σ = Stefan's constant [$5.670 * 10^{-8} Wm^{-2}K^{-4}$] A = Area of contact [m^2](s) T = Temperature [K] e = Emissivity of the black body (no units).
$\Delta T = \frac{(I_{in} - I_{out})\Delta t}{C_s}$	T = Temperature [K](v) I_{in} = Incoming radiation intensity [cd](v) I_{out} = Outgoing radiation intensity [cd](v) t = Time [s](s) C_s = Surface heat capacity [Jm^{-2}](s)

TOPIC 9: Motions in Fields

PROJECTILE MOTION	For projectile motion, vertical and horizontal components of velocity can be separated, and if air resistance is ignored, you can treat the action as a parabola, and hence calculate for half-parabola and double your values in order to simplify calculations.
Object Fired Horizontally	
$y = \frac{1}{2}gt^2$ $v = at$ $v = a\sqrt{\frac{2h}{g}}$	y = Height [m] (s) g = Gravitational acceleration [ms^{-2}] (v) t = Time [s](s) a = Acceleration [ms^{-2}](v) h = Height at time of calculation [m](s)

OBJECT FIRED AT AN ANGLE	
$v_t = v_{yo}t - gt$ $y = v_y t - \frac{1}{2}gt^2$ $x = v_{xo}t$	v_y = Current vertical velocity component [ms^{-1}](v) v_{yo} = Initial vertical velocity component [ms^{-1}](v) v_{xo} = Initial horizontal velocity component [ms^{-1}](v) t = Time [s](s) x = Horizontal displacement [m](v)
GRAVITATIONAL FIELDS	
$V_g = \frac{Work\ done}{Test\ mass} = \frac{\Delta E_p}{m}$	V_g = Gravitational potential [Jm^{-1}](v) E_p = Change in gravitational potential energy [J] (s) m = Mass of body in field [kg](s)
$V_g = -\frac{GM}{r}$	G = Gravitational constant [Nm^2kg^{-1}] m = Mass of field-creating body [kg](s) r = Distance from 'M' [m](s)
$g = -\frac{\Delta V}{\Delta r}$	g = Gravitational acceleration or field strength [ms^{-2}] (v) V = Potential in field [Jm^{-1}](v) r = Displacement moved in field [m](s) **Note:** This displacement is only the resultant, regardless of the path taken!
ESCAPE VELOCITY	
g [ms^(-2)] Below surface: $g \propto r$ Above surface: $g \propto \frac{1}{r^2}$ Distance from centre of mass [m] $\frac{1}{2}(mv^2_{escape}) = \frac{GMm}{R_p}$	m = Mass of orbiting body [kg](s) v_{escape} = Escape velocity [ms^{-1}] (v) G = Gravitational Constant [Nm^2kg^{-2}](v) M = Mass of more massive field-creating body [kg](s) m = Mass of orbiting body [kg](s) R_p = Radius of the body of 'M' [m](s) **Note:** $\propto$ means 'proportional to'
ELECTRICAL FIELDS	
$V_g = \frac{\Delta E_p}{q}$	V_g = Electrical potential [JC^{-1}](s) E_p = Electric potential Energy [J] (s) q = Charge on test charge in field. [C] (v)

$V = \dfrac{kq}{r} = \dfrac{q}{4\varepsilon_o r}$	V = Electrical potential [JC^{-1}](s) k = Coulomb's constant [$Nm^2 kg^{-2}$] (s) q = Charge on field-creating particle [C](v) r = Distance from field-creating particle [m](s) ε_o = Permittivity of free space [$C^2 N^{-1} kg^{-2}$] (s)
$E = -\dfrac{\Delta V}{\Delta x}$	E = Electric Field Strength [J](v) V = Electric Potential [Jm^{-1}](v) x = Distance from field-creating body [m](s)

TOPIC 10: Thermal Physics

Work Done By Gas	
$W = p\Delta V$	W = Work done by gas [J](s) p = Pressure of gas(constant) [Pa](s) V = Volume of gas [m^3](s) **Note**: This equation applies for isobaric functions. However, if given a graph, the work done is always the area under the curve
Moving Detector	
$\Delta Q = \Delta U + \Delta W$	Q = Heat [J](s) U = Internal energy [J](s) W = Work done [J](s)
EFFICIENCY	
$e_c = \dfrac{\Delta W}{Q_{hot}} = 1 - \dfrac{Q_c}{Q_H} = 1 - \dfrac{T_c}{T_H}$	e = Efficiency [ratio] W = Work done by engine/ideal gas [J](s) Q_H = Heat taken from hot reservoir [J] (s) Q_c = Heat rejected to cold reservoir [J] (s) T_H = Temperature of hot reservoir [K](s) T_c = Temperature of cold reservoir [K](s)
$e_c = 1 - \dfrac{T_c}{T_H}$	e_c = Carnot Efficiency [ratio] T_H = Temperature of Hot Reservoir [K](s) T_c = Temperature of Cold Reservoir [K](s)

Total Energy of System	
$$U = \left(\frac{3}{2}\right)NkT = \left(\frac{3}{2}\right)nRT = \left(\frac{3}{2}\right)PV$$	U = Internal Energy [J] N = Atomic Number k = Boltzmann's constant [JK^{-1}] P = Pressure [Pa](s) V = Volume [m^3](s) n = Number of moles [mol](s) R = Gas Constant [$K^{-1}mol^{-1}$] T = Temperature [K](s)
Pressure Law	
$$\frac{P}{T} = k$$	P = Pressure [Pa](s) V = Volume [m^3](s) T = Temperature [K](s)
Charles' Law	
$$\frac{V}{T} = k$$	P = Pressure [Pa](s) V = Volume [m^3](s) T = Temperature [K](s)
Charles' Law	
$$PV = k$$	P = Pressure [Pa](s) V = Volume [m^3](s) T = Temperature [K](s)

TOPIC 11: Wave Phenomena

DOPPLER EFFECT	
Moving Source	
$Recieved\ wavelength = \lambda_o = \frac{c - v_s}{f_s}$ $f_o = f_s \left(\frac{1}{1 \pm \frac{v_s}{c}}\right)$	λ_o = Wavelength emitted by source [m] (s) c = Speed of sound (see constants) [ms^{-1}](v) v_s = Velocity of moving source [ms^{-1}](v) f_s = Frequency emitted by source [Hz](s) f_o = Frequency observed by observer [Hz](s)
Moving Detector	
$f_o = f_s + \frac{extra\ waves\ in\ v_o \Delta t}{\Delta t}$ $f_o = f_s \left(1 \pm \frac{v_o}{c}\right)$	f_s = Frequency emitted by source [Hz](s) f_o = Frequency observed by observer [Hz](s) c = Speed of sound (see constants) [ms^{-1}](v) v_o = Velocity of moving detector [ms^{-1}](v)

DIFFRACTION	
Rectangular Aperture	
$\theta = \dfrac{\lambda}{b}$	θ = angle of first minimum (angle) λ = wavelength of incident light $[m]$(s) b = Size of aperture for diffraction $[m]$(s)
Circular Aperture	
$\theta = 1.22\dfrac{\lambda}{b}$ $I = I_o \cos^2(\theta)$ $n = \tan^{-1}(\theta)$	I = Incident intensity [ ](s) I_o = Resultant intensity [ ](s) n = Index ratio θ = Angle of first minimum

TOPIC 12: Electromagnetic Induction

MAGNETIC FIELDS	*Electromotive force comes about whenever 'flux is cut' by a conducting body. It is not actually a 'force', but rather has the same units as potential difference and can be seen as a voltage.*
$B = \dfrac{\Delta \phi}{\Delta A}$ $\phi = BA\cos(\theta)$	B = Magnetic Field Strength $[T]$(v) ϕ = Magnetic Flux $[Tm^2]$(v) A = Area cut by flux $[m^2]$(s) θ = Angle between area enclosed by wires and magnetic field
$\varepsilon = BLv$	ε = Induced emf $[V]$(v) L = Length of area swept out $[m]$(s) v = velocity of moving current-source $[ms^{-1}]$ (v)
$\varepsilon = N\dfrac{\phi}{t}$	ε = Flux linkage or induced e.m.f, applies when there is more than one loop in the coil in which an e.m.f is induced $[V]$(s) N = Number of loops of conductor ϕ = Magnetic Flux $[Tm^2]$(v) t = Time taken $[s]$(s)
TRANSFORMER	
$\dfrac{I_s}{I_p} = \dfrac{V_p}{V_s} = \dfrac{N_p}{N_s}$	I = Current $[A]$(s) V = Voltage $[V]$(v) N = Number of coils around transformer p = Primary loop subscript s = Secondary loop subscript

POWER TRANSMISSION	
$I_{rms} = \dfrac{I_o}{\sqrt{2}}$ $V_{rms} = \dfrac{V_o}{\sqrt{2}}$ $P_{av} = V_{rms} I_{rms} = \dfrac{1}{2} I_o V_o$ $P_{max} = I_o V_o$ $R = \dfrac{V_{rms}}{I_{rms}} = \dfrac{V_o}{I_o}$	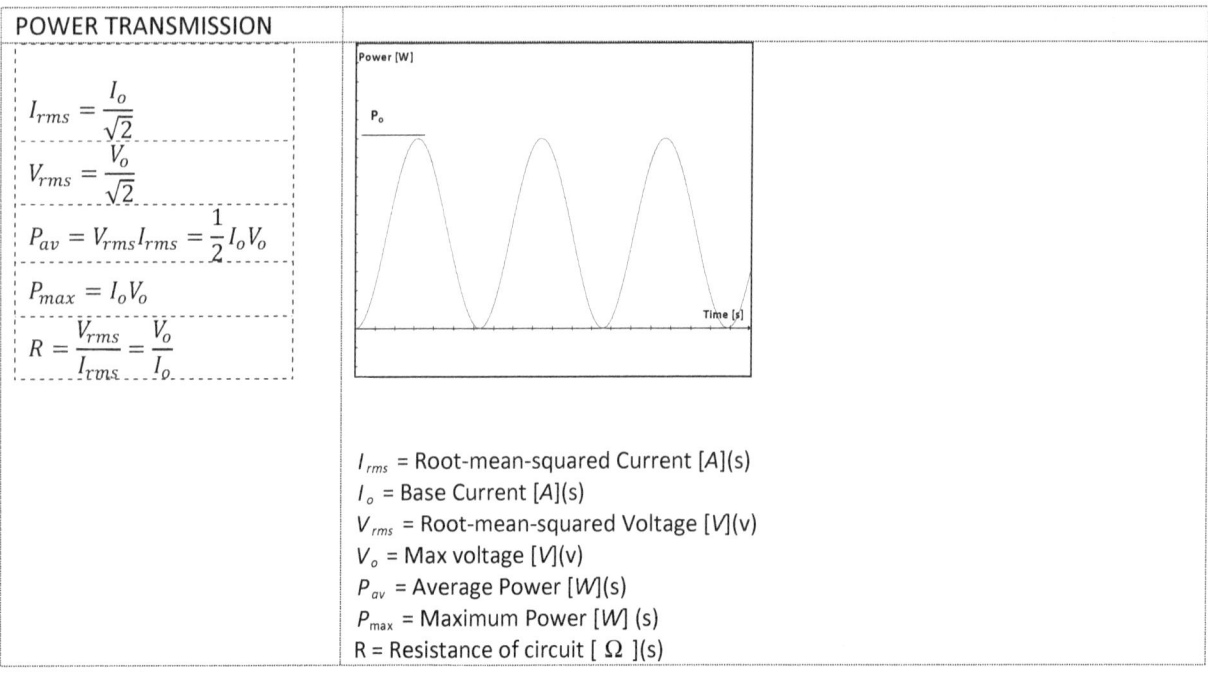 I_{rms} = Root-mean-squared Current [A](s) I_o = Base Current [A](s) V_{rms} = Root-mean-squared Voltage [V](v) V_o = Max voltage [V](v) P_{av} = Average Power [W](s) P_{max} = Maximum Power [W] (s) R = Resistance of circuit [Ω](s)

TOPIC 13: Quantum Physics & Nuclear Physics

PHOTOACTIVITY	
$E = hf$ $hf = \phi + KE = \phi + V_s e$	E = Energy of photon [J](s) h = Planck's constant [Js](s) f = Frequency of photon [Hz](s) ϕ = Work function [Nm^2] (s) KE_{max} = Maximum kinetic energy [Nm^2] (s) V_s = Stopping Voltage [V](v) e = Charge on an electron [C](v)
STOPPING POTENTIAL	
$v = \sqrt{\dfrac{2 V_s e}{m}}$	v = Velocity of photon [ms^{-1}] (v) m = Mass of electron [kg](s) V_s = Stopping Voltage [V](v) e = Charge on an electron [C](v)
$\rho = \dfrac{h}{\lambda}$	ρ = Momentum of photon [$kgms^{-1}$] (v) h = Planck's constant [Js](s) λ = Wavelength of photon [m](s)
$E_k = \dfrac{n^2 h^2}{8 m_e l^2}$	E_k = Kinetic Energy of photon [J](s) n = Energy level (integer value 1,2...) h = Planck's Constant [Js](s) m_e = Electron rest mass [kg] or [u] (s) l = Length of 'box' [m](s)

$\Delta x \Delta p > \dfrac{h}{4\pi}$	x = Uncertainty in the measurement of position [m](s) p = Uncertainty in the measurement of momentum [$kgms^{-1}$](s) h = Planck's constant [Js](s)
$\Delta E \Delta t \geq \dfrac{h}{4\pi}$	E = Uncertainty in the measurement of energy [J](s) t = Uncertainty in the measurement of time [s](s) h = Planck's constant [Js](s)
RADIOACTIVITY/HALF-LIFE	
$N = N_o e^{-\lambda t}$ $A = \dfrac{-\Delta N}{\Delta t} = \lambda N = \lambda N_o e^{-\lambda t}$ $T_{\frac{1}{2}} = \dfrac{\ln(2)}{\lambda}$	N_o = Original number of radioactive atoms at beginning of testing N = Current number of radioactive atoms in a substance λ = Decay constant [s^{-1}](s) t = Time taken [s](s) A = Activity [Bq](s) $T_{\frac{1}{2}}$ = Half-life [s](s)

TOPIC 14: Digital Technology

CAPACITANCE	
$C = \dfrac{Q}{V}$	C = Capacitance [CV^{-1}](s) Q = Charge stored in capacitor [C](s) V = Voltage difference between capacitor plates [V](v)
MAGNIFICATION	
$Magnification = \dfrac{Length\ of\ image\ on\ CCD}{Length\ of\ object}$	
PIT-DEPTH CALCULATION	
$Pit\ depth = \dfrac{\lambda}{4}$	The pit depth equation applies for both CD's and DVD's.
BINARY CONVERSION	
Ex: 9 	8 =(2^3)	4=(2^2)	2=(2^1)	1=(2^0)
---	---	---	---
1	0	0	1		Make a table as on the left, and the summation of the components corresponds to the integer value which the binary number represents.
QUANTUM EFFICIENCY	
$e = \dfrac{No.\ of\ photoelectrons\ emitted}{No.\ of\ photons\ incident\ on\ pixel}$	

NOTES

DEFINITIONS

During IB exams, most questions are heavily structured around an understanding of 'key words' in the syllabus. Each of these terms in this section therefore has been taken straight from the syllabus, with explanations given more colloquially. The dictionary section is broken down according to the syllabus, not alphabetically. Most of these definitions involve a certain amount of assumed knowledge, and where possible, examples are given to put the different variables into context. If you can come to terms with the definitions and apply them accordingly, you're already half way to understanding the course.

Core Syllabus

TOPIC 1: Physics & Physical Measurement

1.1: Range of Magnitudes of quantities in our universe

- **Order of Magnitude:** The size of a value, determined by the number of powers of 10. For example, 1000 can be represented as 10×10^3, where 10^3 is the order of magnitude.

- **Significant Figures:** How many non-zero integer values are used to represent a value. These only count after the first non-zero is presented. Eg. 0.00235 is presented to 3.s.f, as is 2.35. The more significant figures something has, the more accurate it tends to be.

1.2 Measurements & Uncertainties

- **Fundamental Unit:** The SI units that come after the magnitude of a value, to show what it is measuring. Fundamental means it cannot be broken down any further - it is not derived. The form used in the IB is 'SI', meaning International System of Units.

- **Derived Unit:** Like fundamental units, it is what comes after the magnitude of a value, to show what is measuring. Derived units however are units which can be broken down into constituent parts which are fundamental. For example, a Watt [W] is derived and can be broken into its fundamental form $[Nms^{-1}]$.

- **Random Error:** As the name suggests, these come about due to a one-off 'mistake', such as reading a 3 as an 8. These errors can be eliminated by repeated readings.

- **Systematic Error:** An error due to mis-calibration or an error which repeats itself due to an error in the method of reading a value. These errors cannot be eliminated by repeated readings.

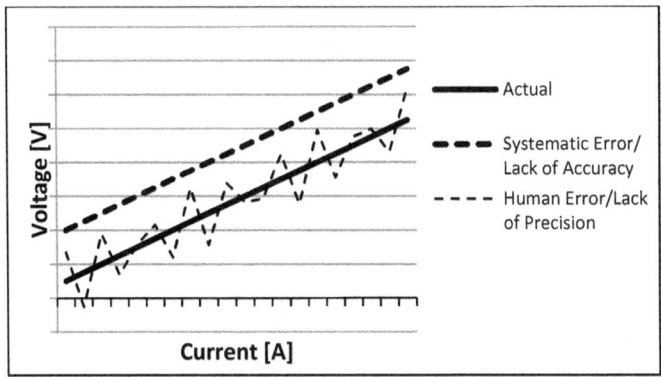
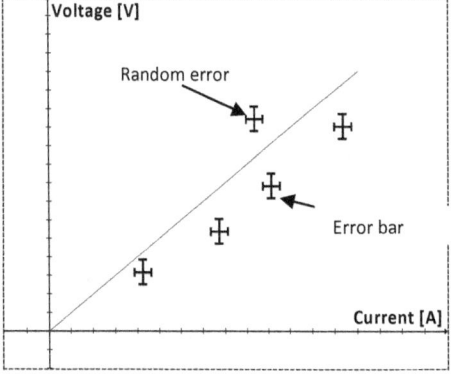

- **Precision:** A precise measurement is one with minimal random error.

- **Accuracy:** An accurate value is one with a small systematic error.

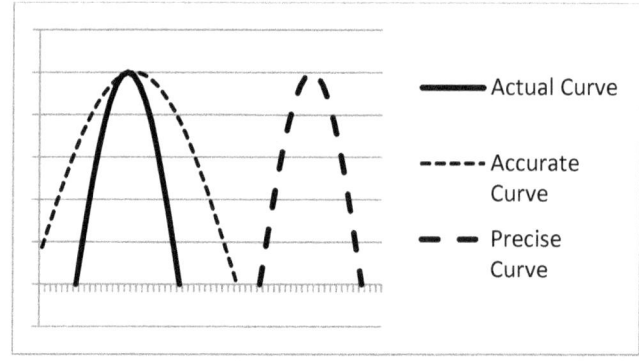

- **Error Bar:** Made up of a horizontal and/or vertical component(s), and shows the uncertainty of each measurement. A line of best fit is usually formed to pass through as many error bars as possible.

- **Vector Quantity:** A quantity with both magnitude and direction. Eg:. velocity.

- **Scalar Quantity:** A quantity with only magnitude. Eg. speed.

TOPIC 2: Mechanics

2.1 Kinematics

- **Displacement:** Distance moved in a particular direction.

- **Speed:** $\frac{Distance}{time}$, how fast something is moving.

- **Velocity:** $\frac{Displacement}{Time}$, how fast something is moving in a particular direction. The rate of change of displacement.

- **Acceleration:** $\frac{Velocity}{Time}$, how fast something's speed is changing with time in a particular direction. The rate of change of acceleration.

- **Instantaneous:** A value taken at a particular time, where time is taken to the limit of 0, $\lim_{\Delta t \to 0}(\Delta t)$.

- **Average:** If considering a linear system, (final value + initial value)/2.

- **Relative Velocity:** How fast something is moving in a particular direction as seen by a particular body, which may be moving as well. Eg. If you are walking in a straight line at 3 ms^{-1} and someone walking next to you at 5 ms^{-1} passes you, their velocity relative to you is 2 ms^{-1}.

2.2 Forces & Dynamics

- **Mass:** Amount of matter possessed by a body.

- **Weight:** A force, resulting from how much mass there is in a body multiplied by the net acceleration on it.

- **Free-body Diagram:** A drawing of a body showing all the forces acting *on* the body. Remember, force is a vector so the arrows showing the forces must show both magnitude (in length of the arrow) and direction.

- **Newton's First Law of Motion:** SEE LAWS

- **Translational Equilibrium:** If the resultant force on a body = 0, it is in translational equilibrium. Translational means in one plane, with the equilibrium existing without rotation. This is simplified to $\sum F = 0$.

- **Newton's Second Law of Motion:** SEE LAWS

- **Linear Momentum (see FORMULAE):** A means of measuring how much inertia something has. It is the product of a body's mass and velocity, provided the velocity is translational.

- **Impulse (see FORMULAE):** The change in momentum of a body, which can be thought of as a quick 'push' or 'pull' which will change the manner in which its moving.

- **Law of conservation of momentum:** SEE LAWS

- **Newton's 3rd law of motion:** SEE LAWS

2.3: Work, Energy & Power

- **Work (see FORMULAE):** Defined as force on a body multiplied by the distance moved in the direction of the force. Importantly, it is a scalar quantity.

- **Energy:** It is the quantity revealing the ability of something to do work, which can be found in various forms throughout the syllabus.

- **Kinetic Energy (see FORMULAE):** The energy associated with the movement of a body.

- **Gravitational Potential Energy (see FORMULAE):** The energy which is 'stored' by the body relative to where it is in a gravitational field, and could be turned into other forms of energy. It is determined by how far away a body is from the field creating body.

- **Principle of Conservation of Energy:** SEE LAWS

- **Elastic Collision:** A collision between 2^+ bodies where no mechanical energy is lost to the surroundings, but energy may be transferred between the bodies.

- **Inelastic collision:** A collision between 2^+ bodies where mechanical energy (eg. heat, sound etc.) is lost to surroundings, and a portion of each body's energy can be transferred to another.

- **Power:** The rate at which energy is transferred. $\frac{Energy\ Transformed}{Time\ Taken}$.

- **Efficiency:** This is a value which has no units, but rather can be seen as a percentage of how 'well' energy has been transformed, in the ratio $\frac{Useful\ work\ out}{Total\ Energy\ Transformed}$. The following must always apply: $0 \leq efficiency \leq 1$.

2.4: Uniform Circular Motion

- **Uniform Circular Motion:** A body moving in a circle at constant speed. Imagine a ball hanging from a string where you push it to move in a circle on the same plane, if there were no air resistance or gravity (eg. in a vacuum), the ball would propagate at constant speed carving out the same circle indefinitely. The velocity component of the ball will always be instantaneously tangential to its position in the circle.

- **Centripetal Acceleration (see FORMULAE):** The acceleration of a body propagating in circular motion. It is always directed to the centre of the circle, and is a vector quantity.

- **Centripetal Force (see FORMULAE):** The force associated with centripetal acceleration, it is directed to the centre of the circle which the moving body carves out, and is equivalent to the centripetal acceleration multiplied by the body's mass.

TOPIC 3: Thermal Physics

3.1 Thermal Concepts

- **Temperature (see FORMULAE):** A measurement of 'hot' and 'cold' which determines which direction heat will flow. It is a linear measurement which can be represented in Kelvin, Celsius or Fahrenheit. It is also the measure of the 'average kinetic energy of a body'.

- **Kelvin Scale (see FORMULAE):** A scale representing variation of temperature. The 'increments' of Kelvin scale are the same as Celsius, but they are offset by 273 'increments' of temperature. $0°K$ represents the state of matter when a body has no energy, and because energy can never be negative, this is what we deem 'absolute zero'. A measure of temperature.

- **Celsius Scale (see FORMULAE):** A scale representing variation of temperature. The same idea as the Kelvin scale which is 273 'increments' higher. Its origin coincides with freezing of water. A means of measuring temperature.

- **Internal Energy:** The addition of the kinetic and potential energy of a body. If an object becomes 'hotter', its internal energy has increased and vice versa.

- **Thermal energy (heat):** Heat refers to the transfer of energy, which will always occur from a cold object to a hot object, unless it is inclined to otherwise by an external energy input (eg. a fridge).

- **Mole:** A mole is a measure of 'how much' there is of something. One mole is defined as the amount of a substance which contains the same number of atoms as 12g of C^{12} (Carbon-12). It is an SI unit.

- **Molar Mass:** This is equivalent to the mass of one mole of a substance.

- **Avagadro's Constant (see constants):** The number of atoms in 12g of C^{12} (Carbon-12).

3.2 Thermal Properties of Matter

- **Heat Capacity (see FORMULAE):** The energy required to raise a body's temperature by 1K.

- **Specific Heat Capacity (see FORMULAE):** The energy required to raise a unit mass of a substance by 1K. Whenever you see 'specific', think 'per unit mass.'

- **Ideal Gas:** A gas which obeys the ideal gas laws for all values of P, V and T. An example which comes very close to this is Helium.

- **Solid:** A substance of fixed volume and fixed shape, where molecules are held together strongly by intermolecular bonds.

- **Liquid:** A substance of fixed volume but varying shape, there are relatively strong forces between molecules, keeping them close together but allowing them to move around. Glass is technically a liquid.

- **Gas:** A substance of varying volume and shape, with very weak forces between molecules. Molecules can be assumed to be independent of one another.

- **Phase Change:** A phase change occurs whenever a substance changes between solid-liquid-gas-plasma, including skipping intermediary steps. This will come as a result of either the input or extraction of energy and will change the bonds between the molecules. On a Pressure-Volume diagram, this will be a horizontal line.

- **Evaporation:** Evaporation is not the same as boiling, in that it occurs for liquids below their boiling point. It occurs at the surface of liquids, where faster moving molecules gain energy from slower moving molecules, enough so to escape as a gas from the liquid. This results in a fall in the overall temperature. Evaporation is a function of surface area, temperature, pressure and wind.

- **Specific Latent Heat:** The amount of energy per unit mass absorbed or released during a phase change.

- **Pressure:** By definition $\frac{Force}{Area}$, which effectively means how many particles are colliding with a wall of the body per given area. The larger the force, the higher the pressure.

- **Fusion:** Phase change from solid to liquid (or vice versa).

- **Vaporization:** Phase change from liquid to gas, (or vice versa).

- **Sublimation:** Phase change from solid to gas/vapour, without any liquid phase (or vice versa).

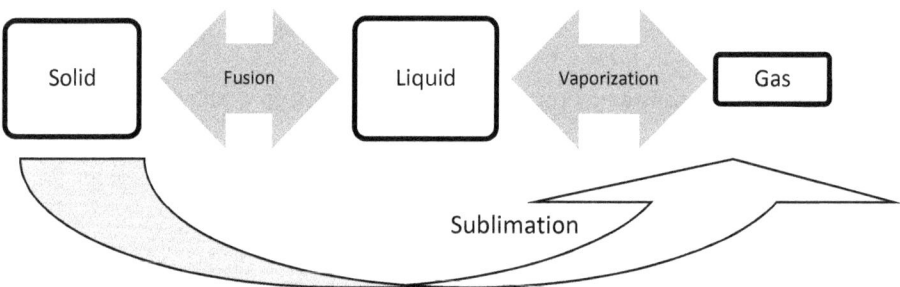

TOPIC 4. Oscillations & Waves

4.1 Kinematics of Simple Harmonic Motion (SHM)

- **Amplitude (A):** The magnitude of a wave, measured from its mean position to the top of a crest or bottom of a trough. The maximum displacement from its mean position.

- **Period (T):** The time taken for one full cycle of a wave to occur. This can be measured from any point on a wave until the identical point is represented again, i.e. trough-to-trough.

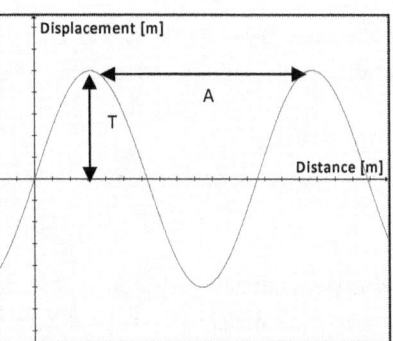

- **Frequency:** The number of oscillations (full waves) which occur per second. It is the inverse of period (see FORMULAE).

- **Wavelength:** The shortest distance on a wave between two successive points in phase with one another.

- **Phase difference:** How many degrees 'out of sync' two waves are. Eg. if two sine-waves are separated by 30°, their phase difference is 30°.

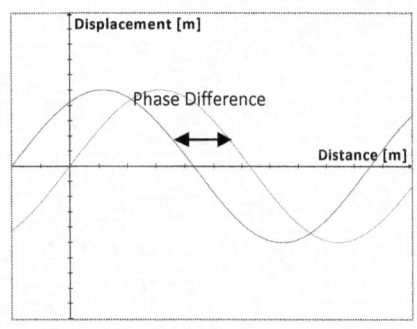

- **Simple Harmonic Motion (see FORMULAE):** The motion of a body subjected to a restoring force directly proportional to the displacement from a fixed point in the line of motion. The resultant waves representing displacement-time, velocity-time and acceleration-time will all be sinusoidal.

4.2/4.3 Forced Oscillations & Resonance

- **Damping:** When energy is lost in a wave due to an external condition or surroundings taking energy away from a wave. This usually comes in the form of friction or viscosity.

- **Under-damped:** When a system has a degree of damping which means the resultant oscillations go on for a long time before reducing to zero.

- **Critically damped:** The minimum degree of damping required to prevent oscillation. It is the degree of damping which results in the displaced body returning to zero displacement without overshooting.

- **Over-damped:** Damping which results in no oscillation. The displaced body slowly returns to zero displacement.

- **Natural Frequency of Vibration:** Otherwise known as 'characteristic frequency', a system vibrates at this frequency when no external conditions are imposed upon a system except for an initial pulse. It is characteristic of the material or body.

- **Forced Oscillations:** When a wave oscillates due to an external driving force, such as a hand creating oscillations in a string.

- **Resonance:** When a wave responds with maximum amplitude to an alternating driving force. This happens when the frequency of the driving force matches the natural frequency of vibration of the wave/substance being driven. A good example is water molecules in a microwave.

4.4 Wave Characteristics

- **Continuous Wave:** Succession of impulses, or succession of individual oscillations.

- **Wave Pulse:** One oscillation that can propagate individually, as seen in diagram.

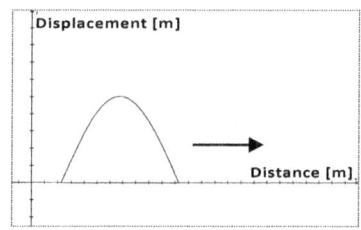

- **Transverse Wave:** Waves where oscillations occur at 90° to the direction of energy transfer, an example of which is the top of a water wave. You can 'see' the wave.

- **Longitudinal Wave:** Waves where oscillations are parallel to the direction of energy transfer, eg. sound waves. This comes about due to a repeating series of compressions and rarefactions of the particles of the medium.

- **Wave-front:** A line drawn to reveal the particles of a wave which are moving together. Eg. all particles on a wave-front will simultaneously be on top of a crest.

- **Ray:** An arrow showing the direction of energy transfer, pointing away from the source of the wave.

- **Crest:** The part of a wave with the largest positive displacement, eg. the top of a wave, which comes about once every period.

- **Trough:** The part of a wave with the largest negative displacement, eg. the bottom of the wave, which comes about once every period.

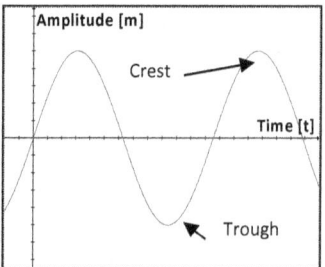

- **Compression:** A point of high pressure in a longitudinal wave, which corresponds to a crest on a pressure – distance graph.

- **Rarefaction:** A point of low pressure in a longitudinal wave, which corresponds to a trough on a pressure-distance graph.

- **Intensity (see FORMULAE):** A measure of the time-averaged energy flux of the wave, derived from multiplying the energy density (energy per unit volume) by the velocity at which the energy is moving.

- **Wave speed (see FORMULAE):** The speed at which a wave propagates through a medium, or the speed at which wave fronts go past a stationary measuring source.

4.5 Wave Properties

- **Incidence/Incident Wave:** The original wave which comes towards a surface.

- **Reflection/Reflected wave:** The wave which bounces off the surface and is 'reflected'. This will have the same speed and wavelength as the incident wave.

- **Transmission/Transmitted Wave:** The wave which goes through the barrier into the second medium. The speed and wavelength of this wave will be different to the incident wave if the two media are different.

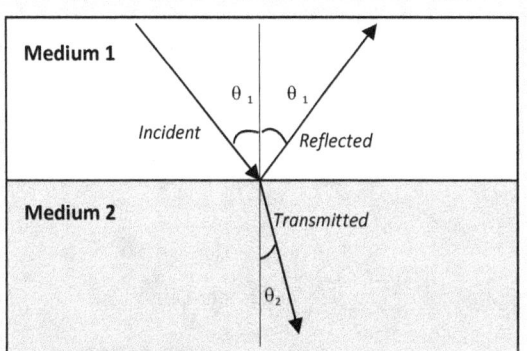

- **Snell's Law:** SEE LAWS

- **Aperture:** A gap between two bodies restricting the movement of a wave which may lead to diffraction.

- **Principle of Superposition:** SEE LAWS

- **Constructive Interference:** When two waves which are in phase combine to give a wave of larger amplitude than their own respective amplitudes.

- **Destructive Interference:** When two waves which are out of phase combine to give a wave of smaller amplitude than their own respective amplitudes.

- **Huygens' Principle:** SEE LAWS

TOPIC 5. Electric Currents

5.1 Electric Potential Difference, current & resistance

- **Electric Potential Difference:** The energy lost by 1Coulomb of charge passing through a 1ohm resistor. Generally, this will be equivalent to 'voltage.'

- **Electronvolt:** A unit of energy, equivalent to one volt times the charge of one electron.

- **Electric Current:** The rate of flow of electric charge. It is essentially the electrons in a conducting lattice which are free to move within the lattice allowing transmission of electricity. Remember that as soon as a circuit is closed and a current flows, electrons all start moving simultaneously throughout the circuit.

- **Resistance: (see FORMULAE):** Defined as the ratio between voltage and current, and can be seen as the components in a circuit which stop or slow down the current by taking away energy from the electrons, or preventing their free movement.

- **Ohm's Law:** SEE LAWS. Something which obeys Ohms law is known as 'ohmic' and will be linear on a V-I diagram. In the plots below, only 'ohmic' represents an ohmic device.

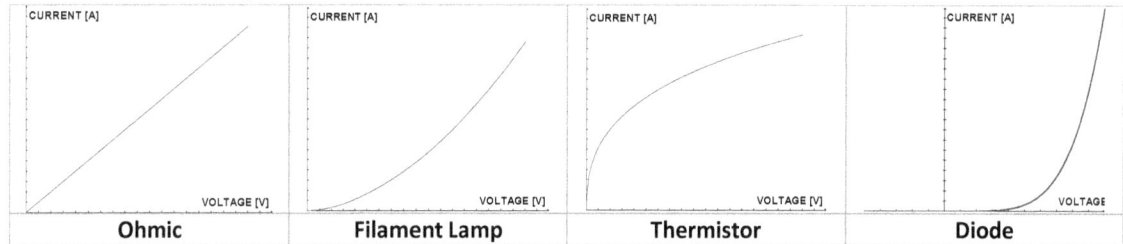

| Ohmic | Filament Lamp | Thermistor | Diode |

5.2 Electric Circuits

- **Electromotive force (emf) (see FORMULAE):** Defined as the energy gained by 1Coulomb of charge passing through a 1V energy source. It is not a force at all, but actually equivalent to the potential difference when no current flows.

- **Internal Resistance:** The resistance present within a power source. Eg. a battery has a certain resistance within it, and this is known as its internal resistance.

- **Ammeter/Ideal Ammeter:** A device which is used to measure the current in a circuit, or a branch within a circuit. An ideal ammeter has zero resistance, and is placed in series with the component you are looking to measure.

- **Voltmeter/Ideal Voltmeter:** A device which is used to measure the voltage between two points, or branch, within a circuit. An ideal voltmeter has infinite resistance, and is placed in parallel with the component you are looking to measure.

א **Potential Divider:** A particular type of circuit consisting of resistors in series to allow a definite fraction of the total voltage to be obtained across a particular component. These systems come up frequently in exam questions. See DERIVATIONS.

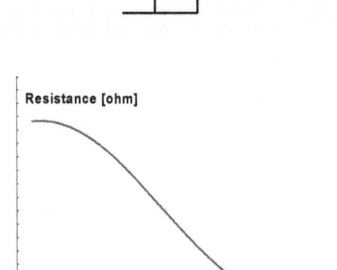

א **Light-dependent resistors (LDR):** Also known as a photoresistor, this is a resistor whose resistance decreases with increasing incident light intensity. An example of when this might be used is a streetlight, since if there is NOT enough light coming into the resistor, it may act as a switch in a circuit to turn the streetlight on.

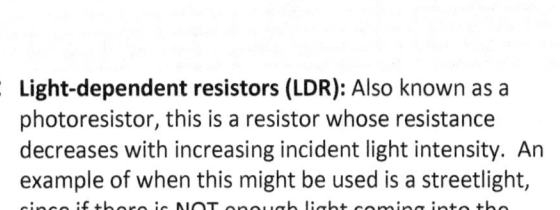

א **Thermistor:** A device made from semiconductor material, whose resistance varies with temperature. When heated, more free electrons are released, making the material a better conductor.

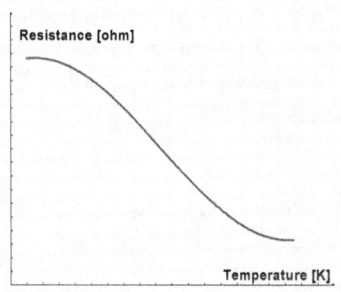

א **Negative Temperature Coefficient (NTC):** The inverse of temperature coefficient, which is a measure of how much a dimension increases with increasing temperature.

א **Strain Gauge:** A device used to measure the strain (= $\frac{Extension}{Original\ Length}$) of a material in terms of the resistance created within an electrical circuit. Used for mechanical purposes.

TOPIC 6: Fields & Forces

6.1: Gravitation Field & Force

- **Newton's Law of Gravitation:** SEE LAWS

- **Gravitational Field Strength:** The force per unit mass in a gravitational field. At the surface of a body, this is equivalent to the acceleration due to gravity on the surface.

- **Test Mass:** A small mass placed within a field in order to test the field strength at a particular point. It will be small enough such that its equal and opposite force on the field-creating body is negligible.

- **Field Line:** A means of describing the state of a field at a particular instance. The further apart they are spaced, the weaker the field. Their direction reveal the state of a field.

- **Equipotential Lines:** Lines which intersect field lines at 90°. They are lines on which all charges (or masses, depending on the type of field) will have equal potential. Their respective spacing reveal the strength of a field. The further apart they are spaced, the weaker the field.

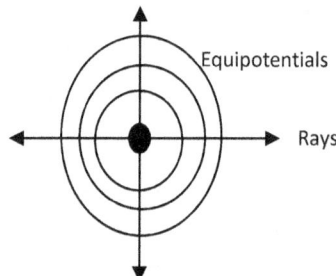

Note: Remember when plotting field lines between two charged plates to include the 'edge-effect', ie bent-lines where the field is weaker at the edges.

6.2 Electric Field & Force

- **Electric Charge:** Either positive or negative. A means of defining which direction charges will interact when near each other. Something without a charge is neutral, as a result of a different number of protons and electrons.

- **Law of Conservation of Charge:** SEE LAWS

- **Conductor:** Any body which allows the flow of charge through it, due to the ability of electrons to move within it to transmit the charge. Eg: steel.

- **Insulator:** Any body which restricts the movement of electric charge. I.e. doesn't have many electrons free to move within a lattice. Eg. wood.

- **Coulomb's Law:** SEE LAWS

- **Electric Field Strength:** The force per positive unit test charge.

- **Test Charge:** A small (positive) charge placed within a field in order to test the field strength. It will be small enough such that its equal and opposite force on the field-creating charge is negligible.

- **Radial Field:** Any field created by a single charge will be radial in that its field lines will propagate outwards in a radial manner, separated by incremental angles.

TOPIC 7: Atomic & Nuclear Physics

7.1: The Atom

- **Bohr Atomic Model:** Also known as the nuclear model, this is the early model of an atom in which there is a small central nucleus surrounded by electrons moving in orbits of defined energy levels.

- **Geiger-Marsden Tube & Experiment:** A set-up used to measure the ionising ability of a substance, in order to determine how radioactive a substance is. Radiation, (for example an alpha-particle) caused by decaying nuclei, ionises gas molecules which in turn cause ions to accelerate finally resulting in an electric pulse. The resultant pulses give us a quantifiable means of telling how radioactive something is.

- **Atomic Energy Levels:** Discrete energy levels associated with electrons orbiting a central nucleus in early atomic models.

- **Emission Spectra:** The spectra of light emitted by an element when it is heated. It is a non-continuous spectrum, only revealing a few characteristic colours. Predominantly black.

- **Absorption Spectra:** When a continuous spectrum of light is shone through an element in its gaseous form, the resultant spectrum will be continuous, barring certain characteristic frequencies which will be absent. These frequencies are exactly the same as the only ones present in the emission spectra of the same element. Predominantly white.

- **Nuclide:** A particular species of atom whose nuclei have the same number of protons, and use the same chemical symbol. Eg. He_2^4 and He_2^5 are isotopes of the same nuclide.

- **Isotope:** Types of nuclide which have the same number of protons, but varying numbers of neutrons. Isotopes have different physical properties due to varying number of neutrons, but have the same chemical properties due to having the same number of electrons.

- **Nucleon/Nucleon number:** A single proton or neutron. Equivalent to *mass number*.

- **Proton Number:** The number of protons in a nucleus. Equivalent to atomic number.

- **Neutron Number:** The number of *neutrons* in a *nuclide*.

- **Coulomb Interaction (see FORMULAE):** The force between any charged particles, as a result of their electric fields. Like forces repel and opposite forces attract.

- **Strong Interaction:** One of the 4 fundamental interactions, it is the force that binds protons and neutrons together in a nucleus.

7.2 Radioactive Decay

- **Natural Radioactive Decay:** A random process which is unaffected by applied conditions. The process whereby the nuclei of some nuclides spontaneously decay, giving off one or more of alpha, beta and gamma particles. How frequently these disintegrations occur is called the activity of the substance, which is measured in Bq.

- **Alpha particle & radiation:** Radioactive decay in which the atomic nucleus emits an alpha particle, which is equivalent to a helium nucleus He_2^4. It is the weakest and slowest of the three forms of radioactive decay, and alpha particles are easily deflected when near other positive charges. It is an ionising form of radiation.

- **Beta particle & radiation:** Radioactive decay where a beta particle is emitted, which is equivalent to an electron β_{-1}^0. The resultant beta particle behaves like a negative charge, has medium penetration ability and variable speed.

- **Gamma particle & radiation:** Radioactive decay in which gamma rays are emitted, which is a form of electromagnetic radiation. It is the most penetrating form of radiation. Gamma rays are undeflected by positive charges, move at the speed of light and have very high penetration ability. After their emission, a nucleus has the same number of nuclides, but less energy. It is effectively the release of energy. γ_o^o

- **Positron:** The *antimatter* or antiparticle version of an electron. It has the same mass as an electron and an electric charge of +1.

- **Neutrino/Antineutrino:** A neutral elementary particle (a lepton), and its antimatter equivalent, which are created as a result of radioactive decay.

- **Radioactive half-life:** The time in which the amount of a radioactive substance's nuclides decay to half its original value. An exponential function. The same statement can be said using the activity of a substance in place of its radioactive nuclides.

7.3 Nuclear Reactions, Fission & Fusion

- **Artificial transmutation:** When you bombard a nucleus with any positive particle in order to create nuclear reactions. Not a naturally occurring phenomenon.

- **Unified Atomic Mass Unit (u) (see constants):** 1/12 the mass of a C-12 atom.

- **Einstein Mass-Energy Equivalence Relationship:** SEE LAWS

- **Mass Defect:** The difference between the mass of a nucleus and the mass of its constituent nucleons, which is found in the form of energy. For example, if you weighed the mass of a helium nucleus, this will be less than the mass of two protons and two neutrons existing individually.

- **Binding Energy:** The amount of energy released when a nucleus is assembled from its constituent nucleons. This amount of energy is equivalent to the mass difference known as mass defect. These two concepts are linked by the mass-energy relation $E = mc^2$.

- **Nuclear Fission:** A nuclear reaction where large nuclei are induced to break up into smaller nuclei, releasing energy in the process. This is the form of nuclear reaction used in today's power plants, and a very efficient means of energy production. A typical energy source is Uranium-235.

- **Nuclear Fusion:** A nuclear reaction where small nuclei are induced to join together into larger nuclei, releasing energy in the process. Although feasible, no power plants currently use this system. It is the reaction that occurs on the sun.

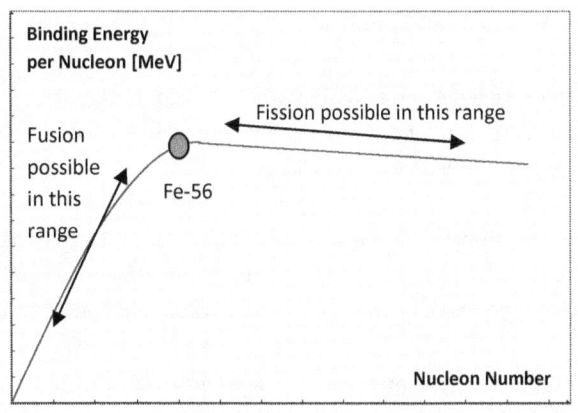

TOPIC 8: Energy, Power & Climate Change
8.1: Energy Degradation & Power Generation

- **Degraded Energy**: Energy which has undergone a transformation into a new form which is less able to do work.

- **Sankey Diagrams/Energy-Flow Diagrams:** Flow diagrams where the width of the arrow is proportional to the quantity of flow. They give a good visual interpretation of energy flow in a process or cyclic process.

8.2 World Energy Sources

- **Renewable Energy Sources:** An energy source which can never be used up. Eg. Wind.

- **Non-renewable energy sources:** An energy source which is definite, and can be used up. Eg. Fossil fuels.

- **Energy Density:** Amount of energy released relative to the amount of source put into the process.

8.3 Fossil Fuel Energy Production

- **Fossil Fuel:** Coal, oil and gas. Fuels which have been created by very high temperatures and pressure beneath the earth's surface. Their origins lie in dead matter, based on the composition of carbon.

8.4 Non fossil-fuel power production

- **Chain Reaction:** This is feasible in fission reactions, where an emitted nucleus goes on to collide with another atom and maintain the process of fission without needing further energy input. However, this can be dangerous if not kept under control.

- **Controlled Nuclear Fission:** A fission process where the rate at which reactions are occurring is controlled so that there is no risk of failure of the system, overheating of components or explosion. This is done via the use of moderators and control rods.

- **Uncontrolled Nuclear Fission:** Uncontrolled nuclear fission occurs when there is no presence of moderators or control rods in order to control the rate at which fission is occurring, a prime example of which is a nuclear bomb which is a very dangerous weapon.

- **Fuel Enrichment:** Changing a fuel so that it is more able to do work. A prime example is giving nuclear fuel (eg. uranium) a higher composition of uranium-235 via isotope separation. It results in a more radioactive fuel-source which can result in faster fission and fusion reactions, providing more energy.

- **Energy Transformations**: The process whereby energy stored as one source is 'used' in order to produce a different type of energy. An example is burning fossil fuels to heat up water which will then spin gas turbines to create electricity. Thus, energy has transformed from one form to another, but conserved throughout, and abides by the law of conservation of energy.

- **Moderator:** A medium in a nuclear power plant which reduces the speed of fast neutrons. The moderator will usually be either water, heavy water or graphite.

- **Control Rods:** Rods in a nuclear power plant which absorb neutrons in order to control the reaction. They are typically made from silver or idium.

- **Heat Exchanger:** A chamber which allows a nuclear reaction to take place without interfering with the surroundings.

- **Nuclear Waste:** The waste product of a nuclear reaction, in the form of radioactive material. A highly controversial matter, nuclear waste will come in the form of the original power source (eg. uranium) which was initially ore and can no longer be used in the reaction.

- **Photovoltaic Cell:** A semiconductor panel which converts radiated heat into a voltage which can then be used as a power supply.

- **Solar Heating Panel:** A panel containing a pipe with water (or another fluid), which reflects as much radiated sun energy into the water in order to heat it.

- **Hydroelectric power:** Power generated by the movement of water.

- **Wind Generator:** A device which generates electrical power from wind energy, usually consisting of turbines mounted to an electrical generator.

- **Oscillating Water Column (OWC):** A rising and falling water surface produces an oscillating air current, which is then turned into electricity, which is used as a source of energy.

8.5 Greenhouse Effect

- **Albedo:** The fraction of incident light diffusely reflected from a surface. Eg. how strongly an object reflects light from a light source shining on it. Therefore, it's a means of showing something's reflectivity.

- **Greenhouse effect:** An effect caused by an atmosphere containing gases which absorb and emit infrared radiation, by trapping heat within the atmosphere, heating the planet to heat. The key example is the sun's radiation, which is allowed through the ozone. The energy which is not absorbed by the earth will reflect back, but in turn will reflect back off the ozone without being let out again.

- **Greenhouse gases:** Consisting of water vapour, carbon dioxide, methane and ozone, which all contribute to the greenhouse effect by locking in infrared radiation.

- **Black-body radiation/Thermal radiation:** Electromagnetic radiation emitted from the surface of an object due to its temperature.

- **Stefan-Boltzmann Law:** SEE LAWS

- **Emissivity:** The relative ability of a material's surface to emit energy by the process of radiation. The equation for emissivity is: $\frac{Energy\ radiated\ by\ material}{Energy\ radiated\ by\ black\ body\ at\ same\ temperature}$.

8.6 Global Warming

- **Global Warming:** The increase of the earth's average air and ocean temperatures, which supposedly has been occurring since the mid 20th-century. It is a highly contentious issue, and one which many scientists and academics continue to debate.

- **Enhanced Greenhouse effect:** Greenhouse effects which are caused directly by human activities.

- **Coefficient of volume expansion:** A coefficient describing how an object's size or volume changes with temperature. It measures the fractional change in volume per degree change in temperature at constant pressure. It is a material property.

- **Combined heating and power systems (CHP):** Using a heat or power station to generate both electricity and useful heat at the same time. It is a form of energy 'recycling'.

- **Carbon capture and storage (CCS):** A process of capturing carbon dioxide from places such as fossil fuel plants and storing it in various safe locations in order to combat climate change and global warming.

- **Hybrid Vehicles:** A vehicle which uses two power sources to operate. The most common form is combining an internal combustion engine with an electric motor in order to cut down the use of fossil fuels for petrol.

- **Kyoto Protocol:** A UN protocol which targets combating climate change and global warming. There are currently 187 nations abiding to the protocol. The aim of the protocol is to reduce emissions of the 4 greenhouse gases and two associated gases (hydrofluorocarbons and perfluorocarbons).

Higher Level Syllabus

TOPIC 9: Motion in Fields

9.1: Projectile Motion

- **Projectile Motion:** Motion of a body in a parabolic arc which has both a vertical and horizontal component of velocity. The key forces acting on such a body is gravity, which is constant.

- **Air resistance:** Friction caused by the air which removes energy from a moving body. It affects both the vertical and horizontal components of a projectile.

9.2 Gravitation Field, Potential & Energy

- **Gravitational Potential:** A counter-intuitive concept, whereby the potential is taken to be 0 infinitely far away from the body (because no force acts here), and becomes more and more negative (hence smaller) as you get closer to the source of the field. However, for very small radii (for example, a few metres above the earth), where 'g' can be assumed constant, PE = mgh applies.

- **Gravitational Potential Energy:** The potential energy of a body derived from its position in a gravitational field. The energy which is 'stored' by the body relative to where it is in a gravitational field, and could be turned into other forms of energy. It is determined by how far away a body is from the field creating body.

- **Gravitational Field Line:** A line propagating away from the source of the field, showing the direction in which the gravitational force acts within that field. The further apart these lines are spaced, the weaker the field.

- **Gravitational Equipotential:** A line/curve representing positions within a gravitational field with equivalent gravitational potentials. These lines will always intersect field lines at right angles.

- **Escape Speed:** The speed of an orbiting body within a gravitational field required to escape its orbit and either move into a further orbit, or away from the field entirely. Derived by equating kinetic energy of the orbiting body with the difference in energy between its gravitational potential energy at the surface of the field-creating body and infinity.

9.3: Electric Field, Potential & Energy

- **Electric Potential:** A counter-intuitive concept, whereby the potential is taken to be 0 infinitely far from the field-creating body (because no force acts here), and becomes more and more negative (hence smaller) as you get closer to the source of the field.

- **Electric Potential Energy:** The potential energy of a charge which is derived from its position in an electric field. The energy is 'stored' by the body relative to where it is in the field, and could be turned into other forms of energy. It is determined by how far away a body is from the field-creating charge.

- **Electric Field Lines:** A line propagating away from the source of the field, showing the direction in which the gravitational force acts within that field. The further apart these lines are spaced, the weaker the field at that point.

- **Electric Equipotential Lines:** A line/curve representing positions within a gravitational field with equivalent gravitational potentials. These lines will always intersect field lines at right angles.

9.4: Orbital Motion

- **Kepler's 3rd Law:** SEE LAWS

- **Total energy of satellite (see FORMULAE):** Kinetic Energy + Gravitational Potential Energy

- **Weightlessness:** A condition whereby a body is moving in a gravitational field and the object it is moving within are accelerating in the orbit with identical accelerations. If you then placed a scale under the body, it would give a reading of 0 because the body and the scale have no 'relative' acceleration, and the body is deemed 'weightless'.

TOPIC 10: Thermal Physics

10.1 Thermodynamics

- **Ideal gas:** A gas which obeys the ideal gas law ($PV = nRT$), at all pressures, volumes and temperatures. Assumptions for ideal gases include perfect spheres for its particles, perfectly elastic collisions between particles, no intermolecular forces, no time spent in collisions of molecules, and that all molecules are in random motion.

- **Real gas:** A real gas will not obey the gas law ($PV = nRT$), at certain values of pressure, volume and temperature, most specifically very high and very low temperatures. Furthermore, the majority of the assumptions made about the ideal gas cannot necessarily be assumed, depending on the circumstances.

- **Absolute zero:** Zero degrees Kelvin. The point at which matter ceases to have any form of energy.

10.2 Processes

- **First law of thermodynamics** SEE LAWS

- **Surroundings:** The surrounding fluid which allows for energy transfer. A system which is insulated is insulated from its surroundings.

- **Principle of energy conservation:** SEE LAWS

- **Isochoric/Isovolumetric:** Process of increased or decreased pressure where gas maintains a constant volume.

- **Isobaric:** Process of expansion or compression where gas maintains a constant pressure.

- **Isothermal:** Process of expansion or compression where gas maintains a constant temperature.

- **Adiabatic:** Process where there is no thermal energy transfer with surroundings. It happens rapidly, as can be seen by a steeper gradient than isothermal on a p-V diagram.

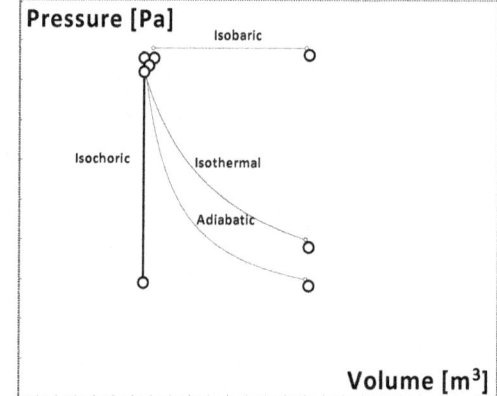

- **Thermodynamic Process:** A process where a gas is taken through variations in temperature, volume or pressure, or a combination of these.

10.3 Second Law of Thermodynamics & Entropy

- **Second law of thermodynamics:** SEE LAWS

- **Entropy:** A property expressing the degree of disorder of a system. Entropy in the universe is always increasing, but in a system, entropy can decrease if the disorder is 'given' to the surroundings.

TOPIC 11: Wave Phenomena

11.1 Standing (stationary) waves

- **Standing Wave:** Occurs when two waves, which are of the same amplitude and frequency and travelling in opposite directions, meet. The resultant wave always seems to 'hold' the same shape, even though it continues to translate. These waves transmit no energy.

- **Node:** Position on a standing wave at equilibrium, or at rest. Always 0 on the displacement axis. Molecules of the medium at this point are compressed.

- **Anti-Node:** Position on a standing wave at maximum displacement, max or min on the displacement axis. Molecules of the medium at this point are 'rarefacted' in a state of expansion.

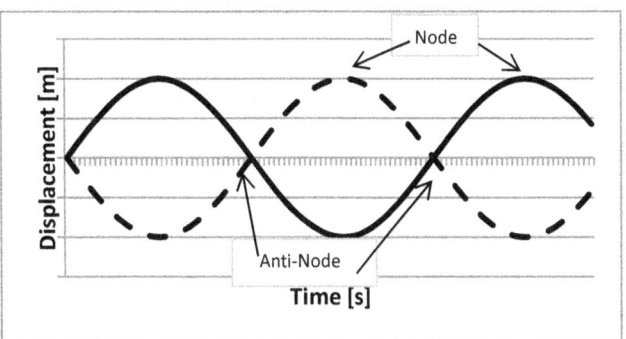

- **Mode of vibration:** These are essentially modes of resonance, which range from 1 to infinity. It reveals the 'frequency' of vibration, which will come in integer multiples of the fundamental mode. The second mode will have twice as many waves 'standing' as the first mode. Modes higher than the fundamental mode are known as harmonics.

- **Fundamental mode/first harmonic:** The mode of vibration with the lowest frequency that can be set up for a given situation. Eg. an organ pipe of length 'L' with one end open and one end closed will have a fundamental mode where the first 'wave' set up obeys: $L = \frac{\pi}{4}$.

- **Travelling wave:** A travelling wave (as opposed to a standing wave), is a wave which transmits energy. This occurs via the particles of the medium which transmit the energy by oscillating about their mean position. Example: a water wave in the sea. Remember that a mechanical wave needs a medium to be transmitted, whereas an EM wave does not.

11.2 Doppler Effect

- **Doppler effect:** An effect created by the combination of a source emitting a wave (typically a sound wave), and either this source moving towards/away from an observer, or an observer moving toward/away from the source. This results in the sound waves either being bunched together or spread apart, resulting in a different observed frequency and wavelength than that which is emitted by the source. Eg: ambulance siren.

- **Shockwave:** A particular instance of the Doppler effect which occurs when the sound-emitting source is moving at the same speed as sound and hence the wave-fronts all bunch up to an infinitely small wavelength. The result sounds like a large bang.

11.3 Diffraction

- **Single-slit experiment:** An experiment where two screens have a small aperture in between them, and a light shone at them. It is then found that the light diffracts after it goes through the aperture. It is a means of showing the diffraction of light, and hence supports the wave-nature of light. The resultant intensity pattern on a screen beyond the aperture is seen in the diagram.

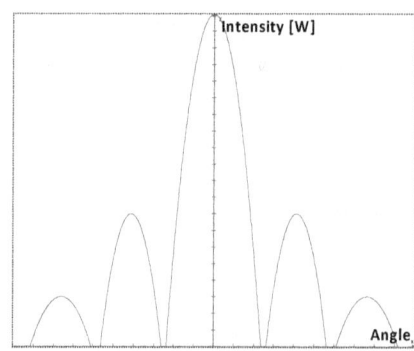

11.4 Resolution

- **Rayleigh Criterion:** SEE LAWS

- **Resolved:** Two sources are deemed to be resolved if one can discern that there are in fact two different sources present.

- **Resolution:** A condition whereby one can tell two light-sources which are similar in frequency to be two distinct sources, based on how much diffraction occurs.

11.5 Polarisation

- **Polarised Light:** Light where the direction of oscillation of the wave is perpendicular to its direction of travel.

- **Brewster's Law:** SEE LAWS

- **Polarizer:** Usually a crystal which is used to produce polarized light travelling in a single plane.

- **Analyser:** Usually a crystal, which is used to show the direction of polarisation of a beam of light. In a standard experiment, the analyser will be placed after the polarizer.

- **Malus' Law:** SEE LAWS

- **Light Intensity (see FORMULAE):** A measure of the concentration of light over a specific area.

- **Optically Active Substance:** A substance which has the ability to rotate the plane of polarization about the direction of motion as it travels through the substance.

- **LCD (Liquid Crystal Display):** A thin panel used for devices such as TVs which uses a polarizing filter to polarize light as it enters and reflective surface at the end of the process to send the image back to the viewer.

TOPIC 12: Electromagnetic Induction

12.1 Induced electromotive force

- **Electromagnetic Force (emf):** Identical to potential difference when no current flows, and is essentially a measure of voltage, and not a force at all. Its definition is the rate at which work is done electrically upon a circuit (power) divided by the current.

- **Magnetic flux:** Essentially how many field lines there are in a magnetic field, and how frequently they are being cut by a conducting body. You can visualise flux as a flow, or water, which flows through a pipe. The larger the cross-section of the pipe, the more water that can flow through, and hence the more flux.

- **Magnetic flux linkage:** Magnetic flux multiplied by the number of windings of the conductor within the magnetic field.

- **Faraday's law:** SEE LAWS

- **Lenz' Law:** SEE LAWS

- **Electromagnetic Induction:** When a conductor cuts the flux of a magnetic field, by either the conductor or the magnets moving, an emf is 'induced' in the conductor by this process. SEE LAWS.

12.2 Alternating Current

- **Alternating Current (AC):** As opposed to DC, this is the kind of electricity that you get through the 'mains' of a building, and comes in the form of a wave.

- **AC generator:** The result of a coil rotating in a magnetic field due to an external force, resulting in a generated AC signal. The work done rotating the coil within the field results in electrical energy.

- **Root Mean Squared (RMS):** AC values divided by $\sqrt{2}$, due to the wave nature of such signals. Also known as the 'rating.'

- **Peak:** The maximum value of the AC wave is known as the 'peak'. Its magnitude is equal for both the positive and negative value.

- **Ohmic/Non-Ohmic:** Impedances which obey/don't obey Ohm's law. I.e. those which produce linear/non-linear plots on V – I plots.

- **Step-up/Step-down transformers:** A combination of two conducting coils wound around a laminated (iron) core. It is used to either step-up (increase) or step-down (decrease) voltage, without changing the frequency. A transformer's operation is based on a principle called *mutual inductance*.

- **(Ideal) Transformer:** A transformer which follows the formula given in the data booklet as accurately as possible. This means that no power is lost in the transformation of the voltage either side of the core.

- **Transmission Lines:** These are the conducting cables used to transmit electricity from one point to another, which ideally would use up no power. In order to do this, electricity is sent at very high currents through cables with as little resistance as possible in order to lose as little power as possible.

TOPIC 13: Quantum Physics and Nuclear Physics
13.1 Quantum Physics

- **Photoelectric effect:** A phenomenon verifying light's existence as particles. When light (*photo*) is shone on metal surfaces, electrons (*electric*) are emitted from the surface as energy - photoelectrons. However, below a certain frequency, which corresponds to how much energy the light has, no photons are emitted. If light were a wave, eventually enough energy would build up and photons would be emitted. Since this is not the case, light must be viewed as particles in this instance.

- **Photon:** An energy 'package'. The amount of energy in a photon corresponds to the energy difference between the levels an electron jumps and consequently either gives or takes the associated photon's energy during the transition.

- **De Broglie Hypothesis (see FORMULAE):** All particles have a 'matter wave' associated with them. This is a means of viewing electrons as both a wave and a particle.

- **Matter wave:** The basis of the De Broglie Hypothesis, a derived concept. All moving bodies have an associated wave, which is known as its matter wave.

- **Wave-particle duality:** The concept that electrons can be viewed both as waves and particles, but never simultaneously. Experiments can be done to 'prove' both of these phenomena.

- **Schrodinger hydrogen atom model:** The 'new' model of an atom, where electrons don't have defined positions, but rather there is a 'cloud' which corresponds to the probability of finding the electron at that point in the atom. Thus, the exact position of electrons is unknown, but the probability of finding it at any location is known.

- **Heisenberg Uncertainty principle:** SEE LAWS

13.2 Nuclear Physics

- **Decay Constant:** A constant of decay which determines how quickly a naturally decaying atom is likely to decay.

- **Bainbridge Mass Spectrometer:** A device used to find the masses of individual nuclei, via the use of magnetic fields, velocity selectors, and the knowledge of isotopes and charges of particles. See DERIVATIONS.

- **Half-life (see FORMULAE):** The time required for a substance's nuclei to decrease by half, or time taken for a substance's activity to halve. This usually applies to radioactive substances, but can be applied to anything which decays.

TOPIC 14: Digital Technology

- **Analogue Signal:** Continuous stream/wave of information, provided between two limits. This is the kind of information that one would get on a cassette, and is easily distorted. You must sample an analogue signal to create a digital signal.

- **Digital Signal:** Takes discrete values, often from an analogue signal and turned into a continuous stream of digital values (0 or 1s) which are then transmitted and read somewhere else. The more values, i.e. the higher the quantization, the more coherent the signal is. The higher the sampling frequency, the more accurate the resultant digital signal.

- **Quantization:** Process of dividing the analogue signal into a digital signal by sampling the analogue signal at regular intervals and giving corresponding quantized values of the signal.

- **Capacitor:** Two conductors separated by a dielectric, which is used to 'store' charge. Such devices are often used to block DC currents whilst allowing AC currents to pass in order to 'smooth' out a signal.

- **Capacitance:** Ratio of amount of charge that can accumulate on the metallic plates / potential difference between the plates of a capacitor.

- **CCD:** Charge coupled device, it is used to obtain images of high resolution in digital imaging, consisting of a silicon chip covered in pixels. They are present in digital cameras.

- **Magnification:** The number of electrons released when light is incident on a pixel, it is proportional to the light intensity.

- **Fidelity:** Similarity between the original signal and the reproduced signal. I.e. how well the resultant digital signal matches the original analogue signal.

- **Perfect Reproduction:** The recording sounds the same regardless how many times it is replayed.

- **ADC:** Analogue-to-digital-converter. Converts a continuous analogue signal into a discrete binary digital signal as a succession of 0s and 1s.

- **DAC:** Digital-to-analogue converter. A device which converts a digital signal into an analogue one, via reverse sampling.

- **Sampling Rate:** How frequently you take samples of an analogue signal to produce a digital one.

NOTES

DERIVATIONS

For most P2 questions, you will need to do some form of derivation, most of which they usually 'lead' you into. If you know a few of the standard derivations before going into your exams, you can guess where they are trying to lead you and understand which steps you may need to take. The majority of these derivations can be done by knowing all the equations in your data booklet very well, but it may be helpful to do repeatedly so that they become second nature. Refer to the FORMULAE section to see what each variable stands for.

Standard Level Syllabus
TOPIC 2: Mechanics
Kinematic Equations

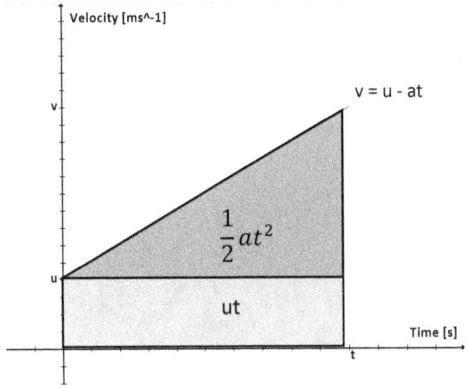

All other relevant equations can be derived from above two, and the plot on the left.

$F = \frac{dp}{dt}$

- $F = ma$

- $a = \frac{\Delta v}{\Delta t}$

- $F = \frac{m \Delta v}{\Delta t}$

- $p = m \Delta v$

- $F = \frac{dp}{dt}$

Derive: v = u + at
- $a = \frac{\Delta velocity}{\Delta time}$
- $a = \frac{v-u}{t}$
- Rearrange for result

Derive: $s = \left(\frac{u+v}{2}\right) * t$
- Average velocity $= \frac{s}{t} = \frac{v+u}{2}$
- Rearrange for result

Derive: $s = ut + \frac{1}{2}at^2$
- Distance = speed x time
- = area under plot
- = Result

TOPIC 3: Thermal Physics
Specific Heat Capacity via Electrical Set-up

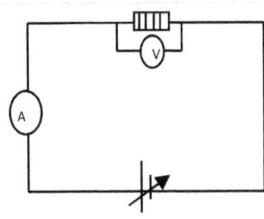

Derive: $c = \frac{VIt}{mc(T_2-T_1)}$
- $c = \frac{Q}{mc\Delta t}$
- $Q = P * t$
- $P = V * I$
- $Q = VIt$
- $c = \frac{VIt}{mc(T_2-T_1)}$

TOPIC 4: Waves
$v = f\lambda$

Derive: $v = f\lambda$
- $velocity = \frac{distance}{time}$
- $v = \frac{s}{t}$
- $f = \frac{1}{T}$
- $v = f\lambda$

Snell's Law

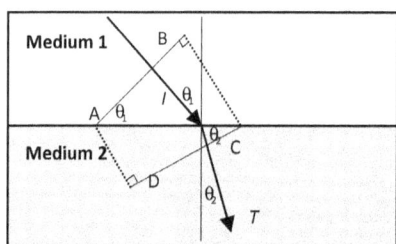

Derive: *Snell's Law*
- Use of Huygen's Principle
- Angles ABC and ADC are 90°
- $\sin(\theta_1) = \frac{v_1 t}{AC}$
- $\sin(\theta_2) = \frac{v_2 t}{AC}$
- $\frac{\sin(\theta_1)}{\sin(\theta_2)} = \frac{v_1}{v_2}$

TOPIC 5: Electricity & Magnetism
Gain in KE when moving within a Potential Difference

Derive: $\Delta PE = Eqd$
- $\Delta Electrical\ PE =$
  $Work\ done\ on\ charge\ in\ an\ electric\ field$
- $W = F * d$
- $F = Eq$
- $W = Eqd$
- $\Delta PE = Eqd$

$I = \frac{Q}{t}$
- $Current = \frac{Charge\ Flowed}{Time\ Taken}$
- $I = \frac{Q}{t}$

Power Dissipation

Derive: $P = VI$
- $Power = \frac{Energy\ difference}{Time}$
- $\frac{Energy\ difference}{Charge\ flowed} * \frac{Charge\ flowed}{Time\ taken} = \frac{Energy\ difference}{Time}$
- $Pd * I = P$
- $P = VI$
- This can be rearranged into other forms but substituting V=IR where relevant to get forms such as $P = \frac{V^2}{R}$

Parallel Resistance

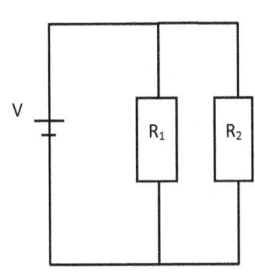

Derive: $\frac{1}{R_t} = \frac{1}{R_1} + \frac{1}{R_2} + \cdots$

- Each leg has the same p.d across it, due to the law of conservation of charge
- $I_t = I_1 + I_2$
- $I = \frac{V}{R}$
- $I_t = \frac{V}{R_1} + \frac{V}{R_2}$
- $\frac{1}{R_t} = \frac{1}{R_1} + \frac{1}{R_2}$

- Above can be extended to $\frac{1}{R_n}$

Potential Divider

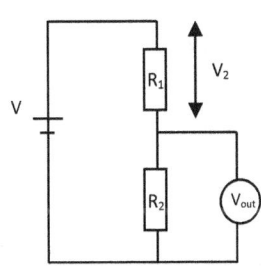

Derive: Potential Divider Fractional Output
- Same current flows through each resistor
- $I_t = I_1 + I_2$
- For whole circuit, V=IR applies, where I is I_t
- $I = \frac{V}{R}$
- $V_{out} = I_2 R_2$
- $V = V_2 + V_{out}$
- $V = I(R_1 + R_2)$
- $\frac{V_{out}}{V} = \frac{IR_2}{I(R_1+R_2)}$

- Thus: $\frac{V_{out}}{V} = \frac{R_2}{R_1+R_2}$

Higher Level Syllabus
TOPIC 9: Fields & Forces
Escape Velocity of Orbiting Body

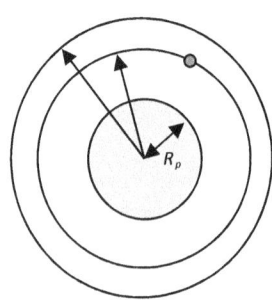

Derive: $v_{escape} = \sqrt{\frac{2GM}{R_p}}$

- Gravitational Potential at planet surface = $-\frac{GM}{R_p}$
- Energy difference for orbiting body b/w planet surface and infinity is: $\frac{GMm}{R_p}$
- Kinetic Energy required to break into further orbit, or escape altogether is at least: $\frac{GMm}{R_p}$
- $\frac{1}{2}(mv_{escape}^2) = \frac{GMm}{R_p}$
- $v_{escape} = \sqrt{\frac{2GM}{R_p}}$

Kepler's 3rd law

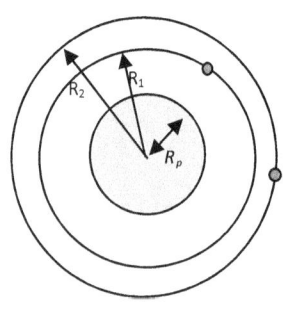

Derive: Kepler's 3rd Law
- For equilibrium of satellite, Gravitational Attraction = Centripetal Force
- Gravitational Attraction: $\frac{GMm}{r^2}$
- Centripetal Force: $\frac{mv^2}{r}$
- $\frac{GMm}{r^2} = \frac{mv^2}{r}$
- $v = \sqrt{\frac{GM}{r}}$
- Satellite has speed $\frac{Orbitting\ Circumference}{Period\ of\ Orbit} = \frac{2\pi r}{T}$
- $GM = \frac{4\pi^2 r^3}{T^2}$
  Which, when substituted above, gives:
- Or: $\frac{r^3}{T^2} = constant$

TOPIC 10: Thermal Physics
Work Done by Gas

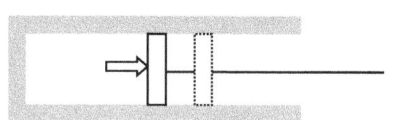

Note: Positive work done corresponds to an increase in volume, which means the gas has done work.

Derive: $W = p\Delta V$
- $W = F * d$
- $p = \frac{F}{A}$
- $F = pA$
- $W = pAd$
- $V = Ad$
- $W = p\Delta V$

Thermal Efficiency

Derive: $\varepsilon = \frac{\Delta W}{Q_{hot}}$

- $Efficiency = \varepsilon = \frac{Work\ done}{Thermal\ Energy\ from\ Hot\ Reservoir}$

- $\varepsilon = \frac{\Delta W}{\Delta t * Power\ from\ Hot\ Reservoir}$

- $Q_{hot} = P_{hot} * \Delta t$

- $\varepsilon = \frac{\Delta W}{Q_{hot}}$

TOPIC 11: Wave Phenomena
Doppler Effect

Moving Source

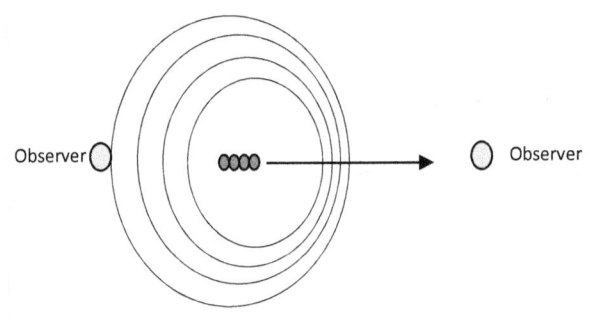

- A moving source results in a change in apparent frequency of sound, as measured by observer
- Wave Equation: $f = \frac{v}{\lambda}$
- If source is moving as constant speed, the centres of the wave-front circles will be equally spaced
- Distance between crests in compressed region:

$$\lambda_o = \lambda - v_s t$$

- But $\lambda = fv$
- Giving: $\lambda_o = \frac{c - v_s}{t}$

- Observed frequency is:
$f_o = \frac{c}{\lambda_o}$
- $f_o = \frac{f_s}{c - v_s} * c$
- $f_o = f_s \left(\frac{1}{1 + \left(\frac{v_s}{c}\right)} \right)$
- The + sign suggests source moving towards the observer. It becomes a '−' sign if moving away.

Moving Observer

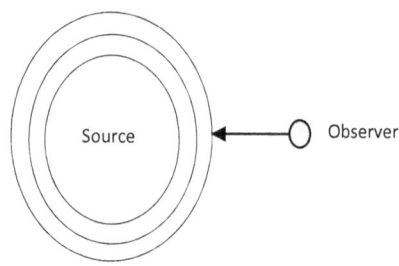

- In time (Δt), observer moves $v_o \Delta t$
- Wave Equation: $f = \frac{v}{\lambda}$
- Explanation
- $f_o = f_s + \frac{v_o t}{\lambda \Delta t}$
- But $\lambda = \frac{c}{f_s}$
- Giving: $f_o = f_s + \frac{v_o}{\frac{c}{f_s}}$
- Observed frequency is:
$$f_o = f_s(1 + \frac{v_o}{c})$$
- The + inside the brackets means the observer is moving towards the source. It becomes a '−' if moving away.

Single-Slit Interference (First minimum)

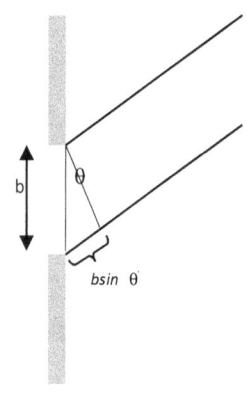

Derive: Brewster's Law
- Using trigonometry, the first minimum is equivalent to the distance mapped on the diagram: $b\sin(\theta) = \lambda$
- $\sin(\theta) = \frac{\lambda}{b}$
- For small angles, can apply assumption: $\sin(\theta) \approx \theta$
- Result is: $\theta = \frac{\lambda}{b}$, which is Brewster's Law!

Brewster's Law

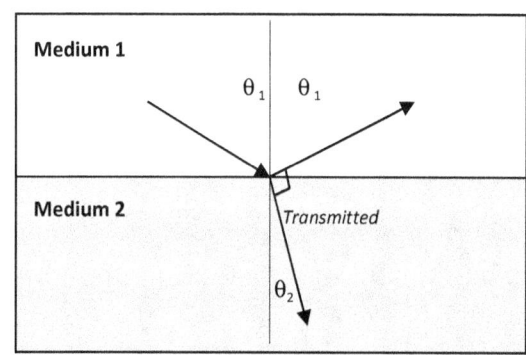

Derive: Brewster's Law
- Reflected and Transmitted at 90°
- From Snell's Law: $n = \frac{\sin(\theta_1)}{\sin(\theta_2)}$
- $\theta_2 = 90 - \theta_1$
- $n = \frac{\sin(\theta_1)}{\sin(90 - \theta_1)}$
- But since: $\sin(90 - \theta_2) = \cos(\theta_2)$
- $n = \tan(\theta_2)$

Double-Slit Experiment

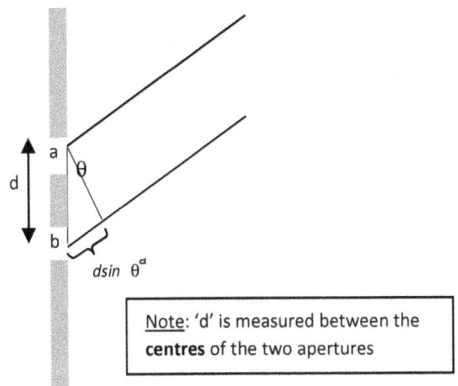

Note: 'd' is measured between the **centres** of the two apertures

Derive: Double-slit Result
- Using trigonometry, path difference is: $d\sin(\theta_2)$
- This will occur again for integer values of wavelength, for n = 1,2,3...
- $d\sin(\theta) = n\lambda$

TOPIC 12: Electromagnetism
EMF induced by relative motion

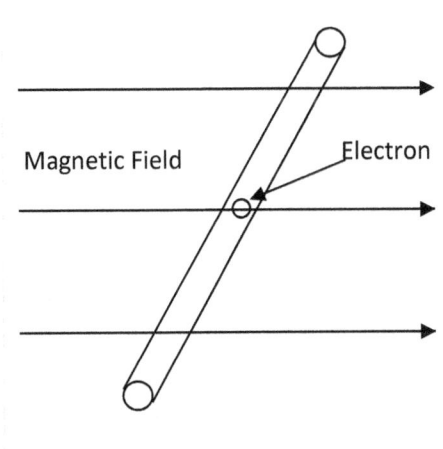

Derive: $emf = Blv$
- Consider electron within a conductor which is inside a magnetic field
- For electron to be in equilibrium:
  Electrical force due to emf = Magnetic force on charge
- Electric Force:
  $Fe = Eq$
- $E = \frac{V}{l}$
- Magnetic Force = Fm
- $Fm = Bqv$
- Equate Magnetic and Electrical Forces:
  $Bqv = \left(\frac{V}{l}\right) * q$
- So, since no current flows:
  $emf = Blv$
- This can then be extended to a whole conductor, where a resultant current would flow.

TOPIC 13: Quantum & Nuclear Physics
Stopping potential experiment

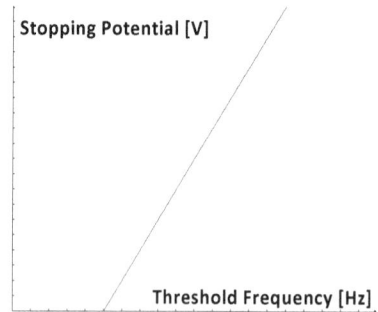

Derive: *Stopping Potential*
- Stopping potential is depends on the frequency of the incident light, and is a measure of the KE of electrons
- $Potential\ Difference \cong V_s = Energy/Charge$
- $KE\ of\ electrons = V_s e$
- $\frac{1}{2}(mv^2) = V_s e$
- $Rearranging, gives$:
$$v = \sqrt{\frac{2V_s e}{m}}$$

Mass spectrometer Isotope Arc Radius

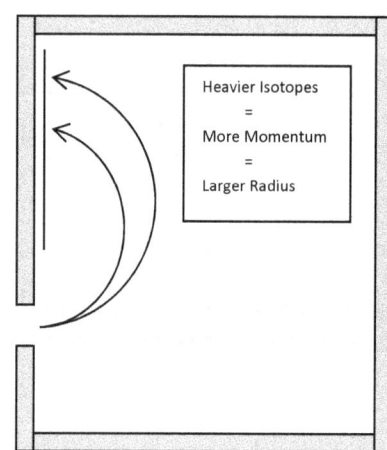

Derive: *Mass spectrometer isotope arc radius*
- Isotopes move in a circular arc, hence can equate: Magnetic Force = Centripetal Force
- $Bqv = \dfrac{mv^2}{r}$
- $r = \dfrac{mv}{Bq}$

Radioactive Half-Life

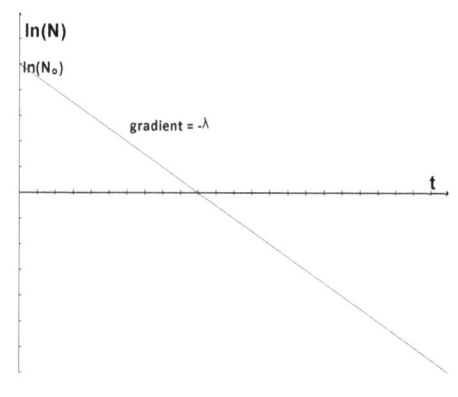

Derive: $T_{\frac{1}{2}} = \ln\left(\dfrac{2}{\lambda}\right)$
- Radioactive Decay Equation $\dfrac{dN}{dt} = -\lambda N$
- Solution to equation: $N = N_o e^{-\lambda t}$
- Taking logarithms of both sides: $\ln(N) = \ln(N_o) - \lambda t \ln(e)$
- $\ln(e) = 1$
- $\ln(N) = \ln(N_o) - \lambda t$
- For half-life $T_{\frac{1}{2}}$
$$N = \frac{N_o}{2}$$
- $\dfrac{N_o}{2} = N_o e^{-\lambda T_{\frac{1}{2}}}$
- $\ln\left(\dfrac{1}{2}\right) = -\lambda T_{\frac{1}{2}}$

- $\lambda T_{\frac{1}{2}} = \ln(2)$
- $T_{\frac{1}{2}} = \ln(2)/\lambda$

NOTES

METHOD & REVISION TECHNIQUES

If you can master a few simple tricks such as quickly re-arranging formulae, you may save yourself a lot of time under exam conditions, as well as being able to second-guess the examiner. Read through these suggestions and apply them the next time you have a test or exam. You may want to read through these quickly before an exam, just so you don't make an unnecessary mistake! I've also included several key facts from the syllabus which are easily skipped over.

Plotting/Graphing Techniques

- When asked to 'sketch' a graph, be sure to include: **a title, axes with name and units, line of best fit (if asked for), error bars (if necessary)** and **points of intersection**, but don't worry about getting axes spacing completely accurate.

- If asked to find the gradient at a particular point of a **non-linear** graph, draw a long tangent at the relevant point and construct a **large** triangle from it. The larger the triangle, the more accurate your result is likely to be. If asked for the gradient of a linear graph, be sure to **use at least half of the total length** for your x and y-values.

- Remember to always plot the **independent** variable on the x-axis and the **dependent** variable on the y-axis.

- Graphical Techniques: It is very important to understand the relations of inverses and where asymptotes exist. Consider the equation $PV = nRT$. What happens if we plot the following, taking RT as a constant (draw for yourself)?
    - P vs. V
    - P vs. T
    - V vs. T

- Consider the equation $y^2 = kx^3$. If asked 'How do we plot y vs. x to get a linear plot?', clearly the answer is y^2 on one axis and x^3 on the other!

- The area under a Force-Time plot is equal to the **impulse**.

- Practise as many cases of displacement-velocity-acceleration as possible, as they are incredibly popular in exams. The two plots on the right should give you a starting point to understanding the principles.

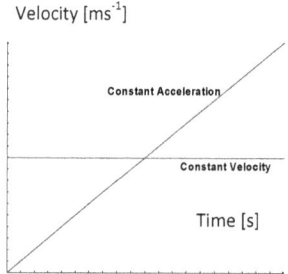
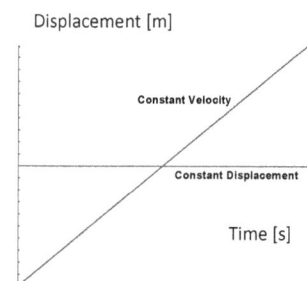

Trigonometry

- A large amount of the syllabus is based on the use of trigonometry. Ensure that whenever using your calculator, you have it set to the correct form between **radians or degrees** before doing calculations.

- For very small angles, you may use the approximation $\sin(\theta) = \tan(\theta)$. It may lead to a cancellation that allows you to re-arrange your equation to fit what they're looking for.

- Vector decomposition is one of the most important techniques to master, as it applies to almost all aspects of the syllabus. Ensure you understand the diagram on the right.

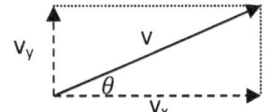

$v_y = v * \cos(\theta)$
$v_x = v * \sin(\theta)$
v = v$_x$ + v$_y$ (vector addition)

- When resolving forces etc. in terms of angles, remember that $\sin(0) = 0$ and $\cos(0) = 1$. If you then imagine something at an angle slightly larger than 0, clearly the larger component will be the cosine.

Mathematical Techniques

- Remember, although there is a difference in more complex mathematics, for IB physics the following calculus forms all mean the same thing: $v^2 = \frac{\Delta y}{\Delta x} = \frac{dy}{dx} = \frac{Change\ in\ y}{Change\ in\ x}$

- For P1 questions, if a question is asking for a ratio, or anything without units, **TRY NOT TO PLUG IN VALUES UNTIL THE END**, as you will probably find that they cancel as soon as you make your ratio and multiplying out will only make your life harder. Keep variables as variables until the last minute and then cancel all that you can at the end.

- ALWAYS **include units** when performing long calculations, since if you're doing a derivation, it may give a tell-tale sign what your next move may be.

- Retain why **work, energy and other quantities are SCALARS**. $W = F\ x\ d$, where d is a VECTOR in the direction of the force applied. When you multiply two vectors, you get a scalar quantity, but remember that for uniform circular motion, the force is always in a different direction to the direction and thus NO WORK IS DONE.

- For Simple Harmonic Motion and Mechanics, you must understand the relationship between displacement, velocity and acceleration. The following is true $a = \frac{dv}{dt} = \frac{d^2s}{dt^2}$. If **displacement 's'** is a sinusoidal function which can be represented by sin(x), clearly its differential (velocity), will be a cosine curve. The differential of velocity (acceleration) will be a negative sin-curve and instantly we've found the relationship between displacement/velocity/acceleration all relative to time.

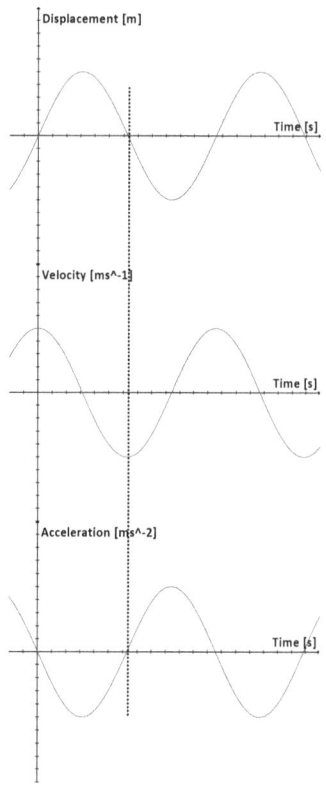

- Occasionally (for projectile motion for example), you will get a result which comes from taking a square root. Given that you get two solutions, sometimes you will need to assume either is possible, but clearly if you get a negative value for displacement if 0 is taken at ground, you must ignore the second value!

- Many P2 questions lead to simultaneous equations. Ensure you know how to solve these problems, and don't be afraid to use your calculator. If using a TI-83 or higher, be sure to download **'Polysymlt'**.

Method & Simplifications

- Always remember to **state any assumptions** when simplifying. For example, when using the kinematic equations, you are assuming constant acceleration, so if the value of acceleration changes, the formulae won't work.

- **Free-Body-Diagrams**: When asked to draw a free-body diagram, there will always be buzz-words in the question. Look out for words like 'frictionless, rough, heavy, uniform', which should immediately tell you what kind of free body-diagram they're looking for. Also remember to **apply Newton's 3rd law**!

- When using **kinematic equations**, remember that they **only work for constant acceleration**. If the acceleration changes, you cannot use them. Also, as long as you clearly state which direction is taken as 'positive' your results will be consistent and you will be given full marks.

- For electrical circuits, if you have a parallel branch, make sure you simplify everything in one loop into a series resistance (i.e. add up the resistances in one loop), and THEN apply the parallel resistance formulae.

- ALWAYS USE KELVIN! Make the conversion between Celsius to Kelvin at the beginning of a question. The simple equation **t(C) = t(K) + 273** MUST be applied at the beginning of a thermodynamics P2 questions in case you refer back and plug in Celsius.

⊕ Physics is based on experimentation, and this is particularly the case for Nuclear and Quantum within the syllabus. Ensure you can replicate all the **experimental diagrams** quickly and thoroughly. The majority of these are included in this revision guide.

Exam Technique

⊕ Make sure you **always write clearly**. Examiners are looking to award marks, so don't write in a hurry so that they can't read your writing and can't give you method marks. Marks are given in stages rather than simply for your final answer!

⊕ **Never write two answers at the end of a question**, in the hope that you will be given credit for one, because examiners won't do that. Show your working and back yourself.

⊕ Always **provide your final answer clearly**, underlined if necessary and always with correct units.

⊕ **Don't skip topics**. Although this may seem a good technique on P2 questions, you will be sure to get caught out on P1 where you need to answer every question.

⊕ **The best tip you will ever receive is to do as many past papers as possible.** If you can learn the common techniques used in exams, and the manner in which they are asked, you've put yourself in a very good position! They should be available either form your school, or online. This will also help build your confidence and pace.

⊕ **When doing past paper questions, never do questions with the answers printed out.** Physics is all about problem solving, and if you can master a technique of problem solving that works for you to get marks, it may be more important than a good grasp of the syllabus. Come exam time, you will have the knowledge that you were able to work through a tricky question previously, so you will be able to do it again!

⊕ **Get your hands on the syllabus.** This guide is based heavily on the syllabus, in order to cover every point. You may find it useful to read through it yourself in order to see where your strengths and weaknesses lie, and even second-guess what questions might be asked.

www.ingramcontent.com/pod-product-compliance
Lightning Source LLC
Chambersburg PA
CBHW051214290426
44109CB00021B/2453